ORGANIC CHEMISTRY
made ridiculously simple

Gene A. Davis, Ph.D.

MEDMASTER

To my parents,
Arthur and Nettie Davis

Contents

PREFACE xi

PART 1. PRINCIPLES OF ATOMIC AND MOLECULAR STRUCTURE AND CHEMICAL
         REACTIVITY 1

CHAPTER 1. Atomic Structure 1

CHAPTER 2. The Octet Rule and Lewis Structures 3
             Ionic Compounds 3
             Covalent Compounds 4

CHAPTER 3. Representing Organic Structural Formulas 7

CHAPTER 4. Covalent Bonds: Energetics and Properties 10

CHAPTER 5. Bond Length and Bond Strength 11

CHAPTER 6. Electronegativity and Bond Polarity 12

CHAPTER 7. Formal Charge 16

CHAPTER 8. Molecular Shapes: Valence Shell Electron Pair Repulsion Theory 18

CHAPTER 9. Atomic Orbitals 21

CHAPTER 10. Orbital Hybridization 23

CHAPTER 11. The Functional Groups 27
             The Hydrocarbons 27
               The Paraffins (Alkanes) 27
               The Olefins (Alkenes) 28
               The Acetylenes (Alkynes) 28
             Aromatic Compounds 29
             Alkyl Halides 30
             Alcohols 30
             Ethers 31
             Carbonyl-Containing Functional Groups 31
               Aldehydes 31
               Ketones 32
               Carboxylic Acids 33
               Carboxylic Acid Derivatives 33
             Amines 34

CHAPTER 12. Physical Properties and Molecular Structure 35

CHAPTER 13. Lewis Acids and Bases 41

Contents

CHAPTER 14. Isomers and Stereochemistry 45
   Constitutional Isomers 45
   Stereoisomers 46
   Optical Isomers 47

CHAPTER 15. Introduction to Organic Reactions 53

CHAPTER 16. Organic Reaction Mechanisms: A General Overview 58
   One-Step Mechanisms 58
   Multi-Step Mechanisms and Reactive Intermediates 59
     Free Radicals 59
     Charged Intermediates 60
       Carbonations (Carbonium Ions) 60
       Carbanions 62
       Other Intermediates 62

CHAPTER 17. Steric Hindrance 64

PART 2. THE FUNCTIONAL GROUPS: PROPERTIES AND REACTIVITY 67

CHAPTER 18. The Alkane Hydrocarbons 67
   Rules of Systematic Nomenclature: Alkanes 69
   Physical Properties of Alkanes 70
   Molecular Shape of the Alkanes 70
   Alkane Chemistry 74
     Preparation of Alkanes 74
     Reactions of Alkanes 75
       Combustion 75
       Halogenation 75

CHAPTER 19. Alkyl Halides: Substitution and Elimination Reactions 76
   Nucleophilic Substitution of Alkyl Halides 76
     The Concerted, One-Step SN2 Mechanism 77
     The Two-Step SN2 Mechanism 77
     Alkyl Halide Structure 78
     Solvent Effects 80
     Steric Hindrance 80
     Nucleophile Concentration 81
     Strength of the Nucleophile 81
     Nature of the Leaving Group) 81
   Elimination Reactions of Alkyl Halides 82

CHAPTER 20. The Unsaturated Hydrocarbons: Alkenes and Alkynes 84
   Reactions of the Alkenes 85
     Addition Reactions 85
       Catalytic Hydrogenation 85
       Addition of Halogen Acids 85
       Addition of Molecular Halogens 85

Contents

Addition of Water 85

Halohydrin Formation 86

Oxidation Reactions 86

Addition/Insertion Oxidation 86

Cis Hydroxylation 86

Epoxidation 86

Oxidative Cleavage Reactions 86

Permanganate Oxidation 86

Ozonolysis 86

Alkene Reaction Mechanisms 87

Carbocation Addition Reactions 87

Addition to Unsymmetrical Alkenes 88

Catalytic Hydrogenation 88

Oxidation Reactions 88

Addition 88

Oxygen Insertion 89

Oxidative Cleavage 89

Oxidative Cleavage Rules 90

Hot Permanganate Oxidation 90

Ozone Oxidation 90

Reactions of Alkynes 91

CHAPTER 21.   Free-Radical Reactions 93

Decomposition of Benzoyl Peroxide 94

Halogenation of Alkanes 95

Anti-Markovnikov Addition of HBr to Alkenes 96

Free-Radical Polymerization 98

CHAPTER 22.   Alcohols and Ethers 100

Structure and Physical Properties 100

Alcohols 101

Synthesis of Alcohols 101

Specific Methods of Alkene Hydration 102

Oxymercuration/ Demercuration 102

Hydroboration 102

The Grignard Reaction 102

Reactions of Alcohols 105

Nucleophilic Substitution 105

Ethers 107

Synthesis of Ethers 107

Reactions of Ethers 107

CHAPTER 23.   Addition and Substitution Reactions of Aldehydes and Ketones 109

CHAPTER 24.   Oxidation and Reduction Reactions of Carbonyl Compounds 112

Reduction Mechanisms (Ketones to 2° Alcohols) 112

Oxidation of Alcohols to Aldehydes and Ketones 115

Contents

CHAPTER 25. Carboxylic Acids and Derivatives 116
Production of Carboxylic Acids 117
Reactions of Carboxylic Acids 118
  Acid-Base Reactions 118
Carboxylic Acid Derivatives 119
  Acid Catalyzed Mechanism 122
  Base Catalyzed Mechanism 122
  Active Esters 123
  Polyamides and Proteins 124
  Reduction 126

CHAPTER 26. Reactions of the Carbonyl Group: A Summary 128

CHAPTER 27. Resonance Structures and Electron Delocalization 131

CHAPTER 28. Aromatic Compounds and Aromaticity 135
Aromaticity and Chemical Reactivity 140

CHAPTER 29. Electrophilic Aromatic Substitutions 141
Mechanism of Electrophilic Aromatic Substitution 141
  General Mechanism of Electrophilic Aromatic Substitution 141
  Specific Reaction Mechanisms of Electrophilic Aromatic Substitution 142
    Halogenation 142
    Nitration 143
    Sulfonation 144
    Alkylation 144
    Acylation 145
    Substitution vs. Addition Reactions 146
    Non-Benzenoid Aromatics 146
Orientation and Reactivity Effects in Substituted Benzenes 148
  Aniline (o,p Director and Rate Activator) 149
  Nitrobenzene (Meta Director and Deactivator) 150

CHAPTER 30. Reactions of Enolate Carbanions 152
Aldol Reaction 154
The Haloform Reaction 155
  The Iodoform Test 156
The Claisen-Schmidt Reaction 157
Claisen Condensation 157
The Dieckmann Reaction 158
Crossed Claisen Reaction 158
Acetoacetic Ester Reaction 159
Knovenagle Reaction 159
Malonic Ester Reaction 160
Micheal Addition 160

CHAPTER 31. The Amines 162
Properties of Amines 163
  Shape 163

Contents

Solubility and Boiling Points 164

Preparation of Amines 166

Hoffman Rearrangement 168

Curtius Rearrangement 168

Reactions of Amines 169

The Hinsberg Test 170

CHAPTER 32. Overview of Reaction Mechanisms: The Production and Fate
of Reactive Intermediates 172

Carbocations (Carbonium Ions) 172

Carbanions 174

Free Radicals 175

One-Step (Concerted) Reactions 176

PART 3. THE SPECTROSCOPIC METHODS OF ANALYSIS 177

CHAPTER 33. Introduction 177

The Electromagnetic Spectrum 177

Spectroscopic Methods and Molecular Structure 178

CHAPTER 34. UV-Visible Spectroscopy 179

CHAPTER 35. Infra-Red Spectroscopy 183

CHAPTER 36. Proton NMR Spectroscopy 186

The Chemical Shift 186

Spin-Spin Splitting 188

Splitting Rules 188

CHAPTER 37. Carbon-13 NMR 191

Chemical Shifts 191

Spin-Spin Splitting 191

Comparison of ^{13}C and ^1H NMR Spectra 192

CHAPTER 38. Mass Spectrometry 193

High Resolution Mass Spectrometry and Molecular Formula 195

REVIEW QUESTIONS 197

GLOSSARY 200

INDEX 204

Preface

The purpose of this book is to help make an undergraduate organic chemistry course successful, easy, and even enjoyable. My approach assumes that organic chemistry is based on a firm foundation of simple and intuitive principles, and that new information can be incorporated, and problems can be solved, by directly applying these basic principles.

My intention is for this book to accompany, not to replace, a standard organic chemistry text. No attempt is made here to include all the details presented in a 1000-page text. The rules of systematic chemical nomenclature are given only brief treatment in this book; uncommon reactions and esoteric mechanisms are mentioned only in passing. Footnoted references, controversial topics, ongoing research, and detailed discussions are left to the standard textbook. The focus here is on basic principles with a few simple examples to offer a smooth and painless entry into each new textbook chapter. Each topic in this study guide is intended to be read and digested before tackling the more detailed expanded textbook version.

Three informational categories are emphasized throughout this study guide. These are **Principles of Chemical Bonding, Functional Group Reactivity,** and **Reaction Mechanisms.**

Principles of Chemical Bonding

The first sections of this book are devoted to a quick review of some key topics from general chemistry. This includes discussions of covalent and ionic bonding, the octet rule, molecular shapes, and bond strength, length, and polarity. The Lewis theory of acids and bases is presented early as one of the bedrock principles of modern chemistry. A discussion of atomic orbitals and orbital hybridization follows. Other principles are discussed as needed at appropriate junctures throughout the book. These include such topics as resonance stabilization via electron delocalization, aromaticity, and steric hindrance, among others.

Functional Group Reactivity

Organic molecules are structures formed of covalently bonded atoms. A few well-defined atomic groupings tend to repeatedly occur in a great many different molecules. These atomic groupings, called *functional groups,* react in much the same ways and confer similar physical properties on all molecules that contain them. Organic compounds contain only about a dozen common functional groups, and each of these engages in only a few important chemical reactions. The reactions and properties of a huge number of organic chemicals can thus be understood by learning the physical properties and chemical reactivity characteristic of these few functional groups. By considering the type, number, and proximity of the functional groups in a given molecular structure, an organic chemist can reliably predict its stability or instability, and its chemical reactivity with many different reagents. He or she can roughly predict the degree of solubility in water and in various organic solvents, as well as other physical properties such as boiling point, melting point, and even odor! Characteristic functional group reactivity and physical behavior greatly simplifies the study of organic chemistry.

Reaction Mechanisms

Reaction mechanisms describe the sequential details of chemical bonds *breaking* in the starting compounds and *forming* to produce the product compounds. Bond breaking steps can produce *reactive intermediates* which are electrically neutral, most often *free radicals,* or which possess a positive or negative charge, called *carbocations* and *carbanions,* respectively. These reactive intermediates are formed in only a few simple ways, and once formed, tend to react in only a few simple and predictable ways. In some reaction mechanisms, bond-breaking and bond-forming occur *simultaneously* in a single step, so no discrete reactive intermediates are formed. This book will emphasize reaction mechanisms whenever possible, since this is the simplest and best way to understand organic chemistry.

Unfortunately, there are some reactions for which the mechanism is not known and others where the mechanism is complicated or unusual. In such cases, reactants and products should simply be memorized. It's a good idea to dedicate a thin notebook, or better, a small pack of index cards, to catalog these reactions for quick recall. It's not difficult. You'll find only a few dozen such reactions for the entire course!

So our approach will be to learn the basic principles of chemical bonding in carbon compounds, and the properties of the resultant molecules; to learn the reactivity and properties of the various functional groups; and to learn the important reaction mechanisms and the production and fate of reactive intermediates. Do these things, and all the details of organic chemistry will easily fall into place!

I express my appreciation for the skillfully rendered graphic art work and helpful comments by Drs. Frank Gorga and Stephen Goldberg, and proofreading by Phyllis Goldenberg.

Gene A. Davis Ph.D.
Lexington, MA
Summer 2002

Part I
PRINCIPLES OF ATOMIC AND MOLECULAR STRUCTURE AND CHEMICAL REACTIVITY

Chapter 1
Atomic Structure

Atoms are the building blocks of all matter. They are composed of nuclear *protons*, each having a positive charge of $+1$, surrounded by a cloud of one or more *electrons*, each with a negative charge of -1. The nuclei of almost all atoms also contain *neutrons*, which have no charge. The simplest and lightest atom is that of the element hydrogen, which is composed of one proton and one electron.

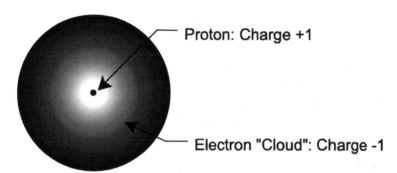

Proton: Charge +1

Electron "Cloud": Charge -1

Fig 1-1: The hydrogen atom

The atoms of the various chemical elements differ from each other based on the number of nuclear protons they contain. The abbreviated periodic table below shows all the chemical elements important in organic chemistry, and some others as well. The atomic number, shown to the lower left of each symbol, identifies the number of protons possessed by an atom of that element.

Group #	I	II	III	IV	V	VI	VII	VIII
	$_1$H							$_2$He
	$_3$Li	$_4$Be	$_5$B	$_6$C	$_7$N	$_8$O	$_9$F	$_{10}$Ne
	$_{11}$Na	$_{12}$Mg	$_{13}$Al	$_{14}$Si	$_{15}$P	$_{16}$S	$_{17}$Cl	$_{18}$Ar
	$_{19}$K	$_{20}$Ca					$_{35}$Br	$_{36}$Kr
	$_{37}$Rb	$_{38}$Sr					$_{53}$I	$_{54}$Xe

Fig 1-2: Abbreviated periodic table.

All atoms have an equal number of electrons and protons, so the algebraic sum of their charges is zero. These uncharged atoms are said to be *electrically neutral*. In this simplified picture of the atom, the electrons are arranged in shells around the nucleus. Atoms of the various elements undergo chemical reactions by gaining, losing, or sharing electrons so as to fill the *outermost* occupied electron shell with a full complement of electrons. This outermost shell is also called the *valence* shell. For the elements which

1

commonly occur in organic compounds, the outer shell is filled by either two electrons (for hydrogen) or eight electrons for the other elements, most notably carbon, nitrogen, oxygen, and the halogens of group VII in the preceding Table. (Sulfur and phosphorous can, however, expand their valence shells to accept more than eight electrons.) Some elements start out with filled outer electron shells and so are chemically unreactive. These elements appear in group VIII of the periodic table and are called the "noble" gases because they are largely impervious to chemical change. *The elements of groups I-VII generally undergo chemical reactions to achieve the stable filled outer electron shells characteristic of the noble gases.*

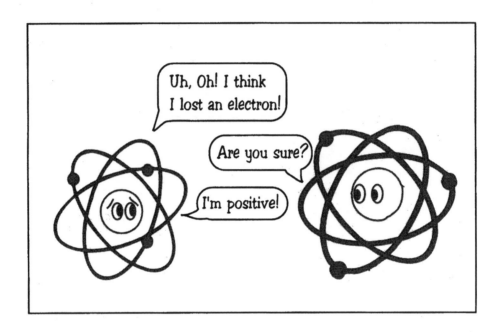

Chapter 2
The Octet Rule and Lewis Structures

The *octet rule* is probably the most important principle of organic chemistry and one of the great insights of modern science. The rule states that *atoms undergo chemical reactions to achieve a noble gas configuration of electrons in their outermost, or valence, electron shell.* Carbon and most of the other elements commonly occurring in organic compounds react to achieve the stable configuration of the noble gases Neon (Ne) or Argon (Ar), which have 8 electrons (an octet) in their valence shells. Hydrogen reacts to achieve the stable configuration of Helium (He), which has two outer shell electrons. *It is only these outer shell or valence electrons which participate in normal chemical reactions.*

The number of valence electrons possessed by a chemical element is indicated by its group position (vertical column) in the periodic table. Excluding various classes of heavy metals, eight such groups exist, usually identified by Roman numerals. Thus, lithium (Li) in group I, has one valence electron, as do the other members of group I, including sodium (Na), and potassium (K). Calcium (Ca) has 2 valence electrons along with barium (Ba), strontium (Sr) and the other members of group II. Likewise, boron (B) has 3 valence electrons; Carbon (C) has 4; nitrogen (N) has 5; oxygen (O) has 6; Fluorine (F) has 7; and Neon (Ne) has 8, since these elements occur in groups III, IV, V, VI ,VII, and VIII, respectively. This is shown in the very abbreviated periodic table above. The elements of group I are often called *alkali metals;* those of group II are called *alkaline earth metals;* and those of group VII are referred to as *halogens.*

Ionic Compounds

To achieve the desired octet, fluorine and the other halogen (group VII) atoms undergo chemical reactions to gain one additional electron. Similarly, oxygen in group VI reacts to gain an additional two electrons, and nitrogen reacts to gain an additional three. In each case these electron gains bring the number of outer (valence) shell electrons up to a total of eight.

$$F \xrightarrow{\text{1e}^-} F^-$$
flourine atom fluoride ion

$$O \xrightarrow{\text{2e}^-} O^{-2}$$
oxygen atom oxide ion

$$N \xrightarrow{\text{3e}^-} N^{-3}$$
nitrogen atom nitride ion

At the left end of the periodic table, lithium, sodium, or potassium atoms in group I will react to *lose* one electron. Calcium and barium, in group II, will tend to lose 2, and aluminum, in group III, loses 3. By electron *loss,* these metallic elements return to an interior shell already filled to a noble gas electron configuration with 2 or 8 electrons:

$$Li \xrightarrow{-1e^-} Li^+$$
lithium atom   lithium ion

$$Ca \xrightarrow{-2e^-} Ca^{+2}$$
calcium atom   calcium ion

$$Al \xrightarrow{-3e^-} Al^{+3}$$
aluminum atom   aluminum ion

These gains and losses of electrons produce oppositely charged *ions* and result in the formation of *ionic compounds,* such as sodium chloride or calcium fluoride:

$$Na + Cl \xrightarrow{e^-} Na^+Cl^- \ (NaCl) \quad \text{Sodium chloride}$$

$$Ca + 2F \xrightarrow{2e^-} Ca^{+2} 2F^- \ (CaF_2) \quad \text{Calcium fluoride}$$

Covalent Compounds

Carbon atoms do not form ionic compounds. To do so would require the *loss* or *gain* of four electrons to achieve the stable helium or neon valence shell configuration. Both of these processes are energetically very difficult to achieve due to the repulsion of like charges. Carbon atoms prefer to acquire an octet by *sharing* their four valence shell electrons with four electrons donated by other atoms. This electron sharing results in the formation of *covalent bonds.* The elements to the right of carbon in the periodic table, found in groups V, VI, and VII, can also form covalent bonds to achieve a stable noble gas electron configuration. Hydrogen atoms can form a single covalent bond to achieve the two electrons of the stable helium(He) configuration.

A carbon atom with its four valence electrons is sometimes denoted as $\cdot\dot{C}\cdot$, with each dot representing a valence (outer shell) electron and the letter C denoting the nucleus and core (inner) electrons. (The nucleus and core electrons do not participate in chemical reactions). The same dot-and-letter symbols, $\colon\ddot{F}\cdot$, $\cdot\ddot{O}\cdot$, $\cdot\dot{N}\cdot$, and H$\cdot$, for example, are used to indicate fluorine, oxygen, nitrogen, and hydrogen, respectively, with the dots explicitly representing the valence electrons of each of those elements.

The reactions below show common examples in which covalent binding can satisfy the octet rule for the participating atoms. In each case, atoms combine to form covalent bonds resulting in *eight valence shell electrons around each reacting atom* (or *two* for hydrogen). Atoms covalently bonded in this way are called *molecules.*

4

	Atoms	Dot Structure	Line Structure	Formula (Name)
Diatomic Gases	2H·	H··H	H-H	H_2 (Hydrogen)
	2:N̈·	:N⋮⋮N:	N≡N	N_2 (Nitrogen)
	2:C̈l·	:C̈l··C̈l:	Cl-Cl	Cl_2 (Chlorine)
Inorganic Compounds	2H· + ·Ö·	:Ö: H H	O H H	H_2O (Water)
	3H· + ·N̈·	N̈ H :: H H	N H H H	NH_3 (Ammonia)
	H· + ·C̈l:	H··C̈l:	H−Cl	HCl (Hydrogen Chloride)
Organic Compounds	4H· + ·C̈·	H H··C··H H	H H−C−H H	CH_4 (Methane)
	4H· + ·C̈· + ·Ö·	H H··C··Ö··H H	H H−C−O−H H	CH_3OH (Methanol)

Fig 2-1: Atoms combining to form covalent molecules by electron sharing.

As shown in the table above, electron pairs can be represented either by a pair of dots or more simply, and more commonly, by a line. Line segments joining two atoms always represent a *two-electron covalent bond*. Pairs of electrons associated with only one atom, and not shared with another, are called *non-bonded pairs*. In the above examples the nitrogen, chlorine, and oxygen atoms have one, three and two non-bonded pairs, respectively. The presence of these can be inferred from the octet rule: the sum of the bonded and non-bonded electrons around these atoms will always equal *eight*. (Non-bonded electron pairs are sometimes omitted for simplicity but are understood to be present.)

Carbon atoms can link to other carbon atoms in many ways to form a wide variety of chemical structures containing different numbers of atoms and different sequences of attachment. This unique property of carbon is called *catenation,* and is responsible for the huge number of different organic compounds as well as the bonding between carbon atoms in the diamond and graphite structures of elemental carbon. Carbon atoms can also link together by both single and multiple bonds to achieve a stable octet of outer shell electrons surrounding each carbon nucleus. *In all cases compliance with the octet rule is mandatory if the molecular structure is to be correctly represented.*

The examples shown above are called *Lewis Molecular Structures.* Writing a Lewis structure which accurately describes covalent molecular bonding requires following only a few simple steps:

1. Write the atomic symbol for each of the participating elements. Indicate the number of valence electrons for each atom by the correct number of dots (which is always equal to the group number of the element in the periodic table).
2. Count the total number of valence electrons contributed by all participating atoms and add these together.
3. Arrange the sum of the valence electrons between the bonded atoms so that all atoms are surrounded by 8 valence electrons, or 2 for Hydrogen. Note that carbon has only bonded (shared) electron pairs; carbon atoms rarely have non-bonded electrons in stable Lewis structures.
4. For simplification, substitute one line for each two-electron bond in the molecule, as shown below.

In the examples below, these four steps are followed to write Lewis structures for some common organic compounds:

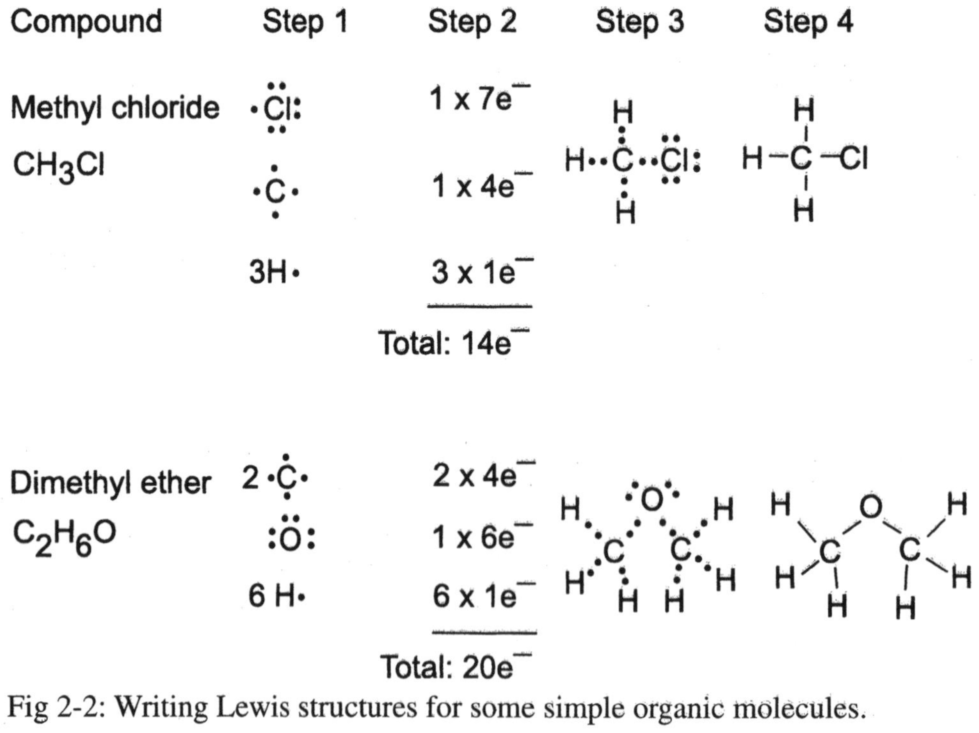

Fig 2-2: Writing Lewis structures for some simple organic molecules.

Chapter 3
Representing Organic Structural Formulas

Lewis structures, for all their simplicity and ingenuity, can at times be unwieldy. Explicitly writing out all the atoms and bonds of a large molecule having, say, over 100 atoms, can be time consuming and difficult. Several simplifications have been invented for this reason, which communicate the same information with greater ease. One of these is called a *condensed* structure, in which atomic groupings are represented with subscripts and parentheses. There is no ambiguity in such structures since compliance with the octet rule mandates only one possible order of connection between the atoms. (If any ambiguity does exist, the groups have been incorrectly represented.) For example, palmitic acid($C_{16}H_{32}O_2$) can be represented by an explicit Lewis structure as:

Or much more simply by a condensed structure as:

$$CH_3(CH_2)_{14}C\begin{array}{c} \diagup O \\ \diagdown OH \end{array}$$

In fact the *carboxyl* group at the right end of the structure can be written more easily as —COOH, since the single and double bonds to the two oxygen atoms are implied by the octet rule and the rules for writing correct Lewis structures.

Similarly, the molecule cyclohexane can be written detailing all covalent bonds or more simply as a condensed structure:

7

Finally, a much more complicated molecule, cholesterol, can be written out fully or by its condensed structure:

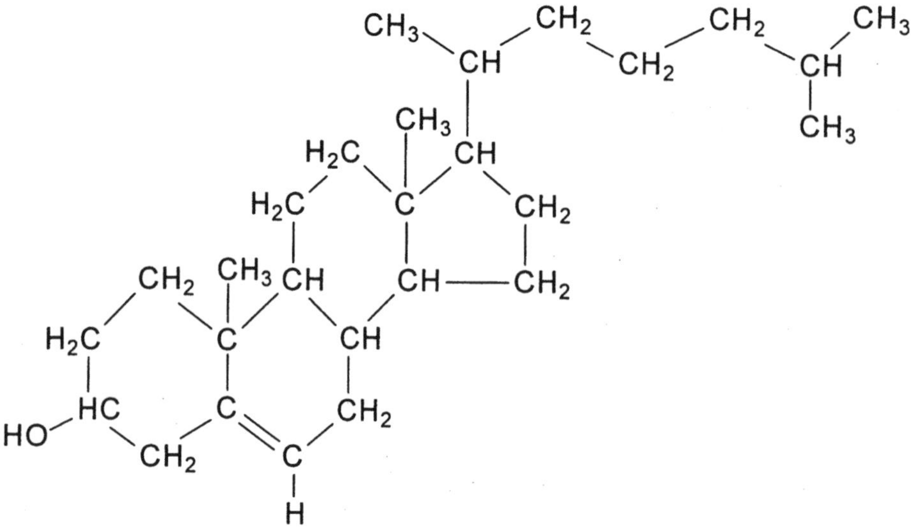

But for a large molecule like cholesterol, which is composed of 74 atoms, even the condensed structure is difficult to write. Another representation called the *bond-line* structure simplifies writing such complex molecules even further. In this method, it is understood that:

1. Each time a line starts, changes direction, or stops, a carbon atom exists at that point and is bonded to the number of hydrogen atoms needed to satisfy the octet rule.

and

2. If atoms other than C and H are present, they are explicitly indicated by the appropriate atomic symbol, for example, O, for the oxygen atom in the cholesterol molecule:

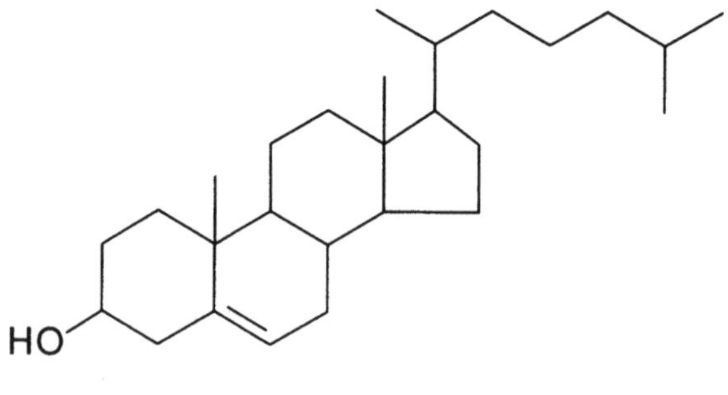

By this simple and ingenious method, complex molecules can be represented unambiguously and with ease. Here are a few more examples of these three methods of representing organic structures:

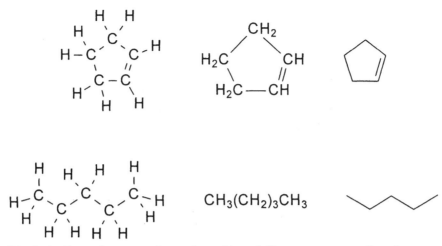

Fig 3-1: Complete, condensed, and bond-line structures of cyclopentene (top) and normal pentane (bottom).

Other, more esoteric, representations also exist which are needed for emphasizing certain geometric features and 3-dimensional shapes of molecules. These will be presented, as needed, throughout the text.

Chapter 4
Covalent Bonds: Energetics and Properties

When two atoms approach each other to form a diatomic molecule, the interaction energy changes as a function of distance between the atoms as shown in the graph below. When far apart (A), each atom is unaware of presence of the other, and their interaction energy is 0. As they approach more closely (B) and each begins to feel the attractive force of the other, the interaction energy assumes negative values as the attractive force between them becomes stronger. At point (C) the two atoms have bonded to form a diatomic molecule. At this point the inter-atomic distance is most favorable; the system is most stable and exists at an energy minimum. The internuclear distance, r_o, is the optimal bond length of the diatomic molecule. Attempt to force the atoms still closer (D) requires the input of a great deal of energy due to a need to overcome the very strong mutual repulsion of the negative electron clouds and of the positive nuclei of the two atoms. Here the interaction energy becomes highly positive. The diagram provides a very useful picture of the energetics of covalent bond formation. Although this discussion refers to the formation of a diatomic molecule, the same considerations apply to the formation of a bond between any two atoms in a polyatomic molecule. The driving force for the formation of the bond from component atoms is, of course, the acquisition of the noble gas configuration of He, Ne, or Ar, having 2 (for helium) or 8 outer shell electrons.

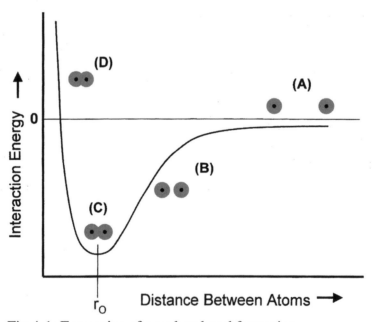

Fig 4-1: Energetics of covalent bond formation.

Chapter 5
Bond Length and Bond Strength

Covalent bonds result from the overlap of electron clouds contributed by the participating atoms. Each such bond has a fixed and characteristic optimal distance between the two bonded nuclei. This distance is called the *bond length*. In addition, covalent bonds between various atoms have characteristic bond energies (or *bond strengths*) as well. This is the energy required to break such a bond, and also the energy liberated when such a bond forms. These strengths vary for each pair of bonded atoms and for the number of bonds linking the atoms together. The bond lengths and bond strengths of covalent bonds commonly occurring in organic molecules are shown in the table below. A few useful generalizations can be made from these tabulated data:

1. Certain single bonds are strong. These have bond energies of about 90 to over 100 kcal/mole.
2. Multiple bonds are stronger and shorter, in aggregate, than single bonds.
3. Certain bonds are especially weak, having bond energies of only 33-57 kcal/mole. These include O-O, C-I, and halogen to halogen bonds, like Br-Br and I-I. Weak bonds break with relative ease, so high chemical reactivity can be expected from this category of covalent bonds.

Bonded Atoms	Length (Å*)	Strength (kcal/mole)	Bonded Atoms	Length (Å)	Strength (kcal/mole)
H-H	0.74	104	O-O	1.32	33
C-C	1.54	83	C=C	1.34	146
C-H	1.10	99	C=N	1.27	147
C-N	1.47	70	C=O	1.22	178
C-O	1.43	84	C≡C	1.21	200
C-F	1.41	105	F-F	1.28	38
C-Cl	1.76	79	Cl-Cl	2.00	58
C-Br	1.91	66	Br-Br	2.28	45
C-I	2.10	57	I-I	2.66	36

* Å (Angstrom) is the first letter of the Swedish alphabet. One Å equals 10^{-8} cm.

Fig 5-1: Table of bond length and strength of common covalent bonds.

Chapter 6
Electronegativity and Bond Polarity

When like atoms, for example two carbon atoms, are covalently joined, the two nuclei exert exactly the same attractive forces on the bonding electron cloud:

Therefore the electron cloud and its negative charge is uniformly distributed between the bonded atoms. Such bonds are said to be *non-polar* and have no excess or deficiency of charge at either end of the bonded atoms. A few simple examples:

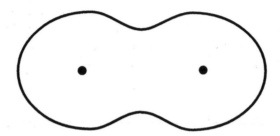

On the other hand, when *unlike* atoms are bonded, one nucleus generally has a greater ability to attract the negatively charged electron cloud than the other. This gives rise to a *partial charge separation* between the atoms resulting in a permanent *bond dipole* and a *polar* covalent bond. The more strongly electron-attracting atom is said to have a greater *electronegativity,* and assumes a partial negative charge, while the less electronegative atom necessarily assumes a partial positive charge. (Partial charges are generally indicated by a lower case Greek letter delta-δ). The bond dipole is often indicated by a crossed arrow, ↦, indicating the positive and negative ends of the dipole at the cross and arrow, respectively. Some examples of polar bonds:

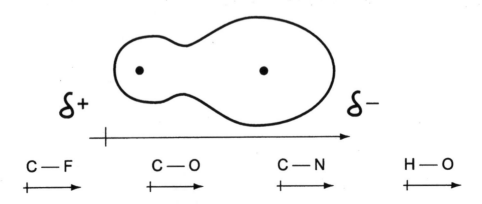

It is worth noting at this point that all chemical bonds, including both ionic and non-ionic, can be described on a continuous scale by the extent of charge separation between neighboring atoms:

Bond Type	Electron Density	Charge Separation	Example
Non-polar covalent		no	Cl—Cl
Polar covalent		yes	H—Cl
Ionic		complete	$Na^+ Cl^-$

Fig 6-1: Charge separation between atoms in various types of ionic and covalent bonds.

Thus, an ionic bond can be viewed as an extreme case of a polar covalent bond in which one of the atoms has assumed all, not just some, of the bonded electron pair.

Shown below is an abbreviated electronegativity table in which numerical electronegativity values are included in a periodic table of the elements. A high electronegativity value indicates a nucleus able to strongly attract covalently bonded electrons.

Group #

I	II	III	IV	V	VI	VII
H (2.1)						
Li (1.0)	Be (1.5)	B (2.0)	C (2.5)	N (3.0)	O (3.5)	F (4.0)
Na (1.0)	Mg (1.2)	Al (1.5)	Si (1.8)	P (2.1)	S (2.5)	Cl (3.0)
K (0.9)	Ca (1.0)					Br (2.8)
Rb (0.9)	Sr (1.0)					I (2.5)

Fig 6-2: Electronegativity values for selected elements.

Clearly, the electronegativity of an element increases from left to right on the table. *The degree of charge separation (polarity) in a given covalent bond is proportional to the magnitude of the difference in electronegativity between the bonded atoms.*

In general:

1. Bonds between metals (groups I and II) and non-metals (groups VI and VII) at the extreme ends of the table are *ionic*. For example:

$$Li^+ \ F^- \qquad Na^+ \ Cl^- \qquad Mg^{+2} \ 2Br^- \qquad Ca^{+2} \ 2I^-$$

2. Bonds between *like* atoms are totally *non-polar*. For example:

$$H-H \qquad C-C \qquad N{\equiv}N \qquad F-F$$

3. Bonds between carbon and hydrogen atoms, as in hydrocarbon organics, have little or no polarity due to the similarity of the electronegativity values for carbon and hydrogen. This lack of polarity in carbon-hydrogen bonds is of primary importance in determining the physical properties and chemical reactivity of organic compounds which contain hydrocarbon structural members.

4. All the elements (other than hydrogen) that are covalently bonded to carbon in organic compounds occur to the *right* of carbon on the periodic table and are thus more electronegative than carbon. These bonds are invariably polarized to some extent with a partial positive charge on the carbon atom and a partial negative charge on the other atom. The *magnitude* of bond polarity increases with the difference in electronegativity between the bonded atoms. The polar character of such bonds is of primary importance in determining physical properties and chemical reactivity.

$$C-F \ > \ C-O \ > \ C-N$$

5. Multiple bonds increase the magnitude of charge separation (bond polarity) compared to single bonds between the same atoms.

$$-C{\equiv}N \ > \ -CH{=}N- \ > \ -CH_2-N\diagup$$

$$\diagdown C{=}O \ > \ -C-O$$

This concept of bond polarity and resulting charge separation is one of the bedrock principles of organic chemistry. When combined with the fact that opposite charges attract and like charges repel, this simple idea allows us to explain and understand a wide range of the physical and chemical properties of organic compounds. The table below summarizes a few of these properties, which we will be visiting again and again as we proceed with our study of organic chemistry.

1. Hydrogen Bonding

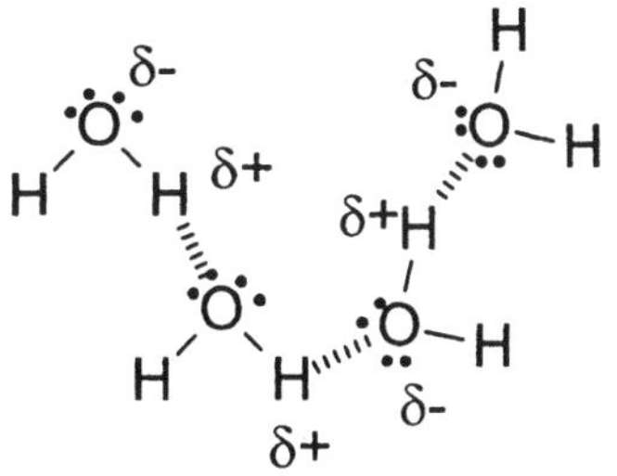

Attractive forces between opposite dipoles cause water molecules to associate by *hydrogen* bonds. This has a major effect on the solvent properties and high boiling point of water.

2. Nucleophilic Reactions

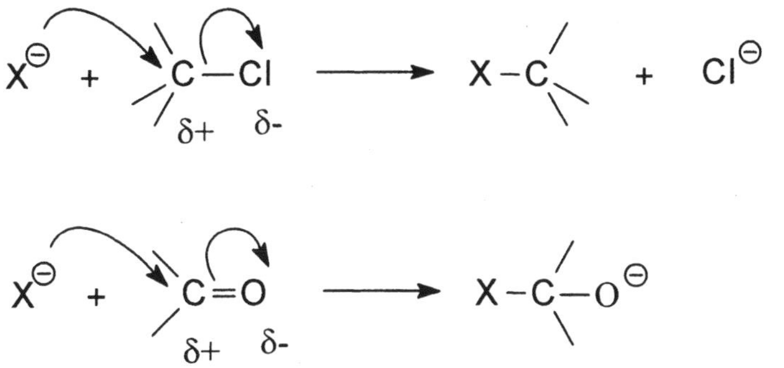

The nucleophile, $X^\ominus$, is attracted to the positively polarized carbon resulting in a substitution or addition reaction.

3. Heterolytic Bond Cleavage

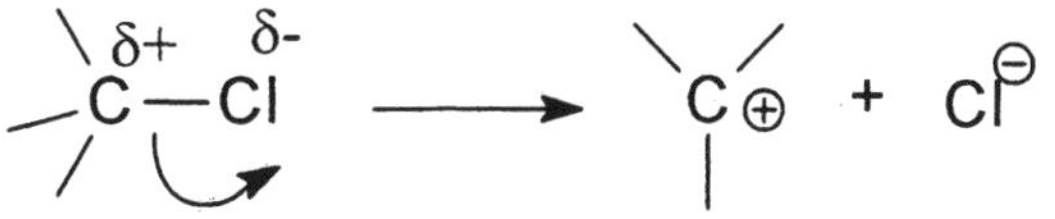

Ion formation by bond cleavage is facilitated by the partial charges on the polar C—Cl bond.

4. Acid Ionization

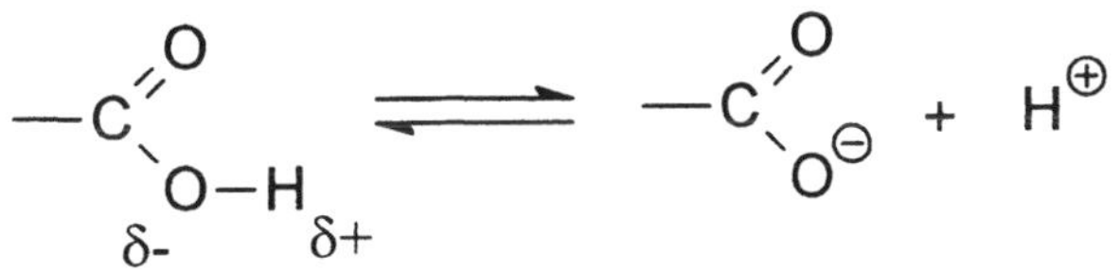

Liberation of the proton is aided by the polar O—H bond of carboxylic acids

Fig 6-3: Some examples of the effect of bond polarity on the physical and chemical behavior of compounds.

Chapter 7
Formal Charge

Covalent molecules can have full charges localized on certain atoms. It is important to identify exactly where these charges reside, since many chemical reactions involve the production or neutralization of charges at specific sites. Keeping careful track of charge is also a book-keeping aid—helping to make sure that a mechanistic pathway is in compliance with the laws of conservation of charge and mass demanded of all physical and chemical processes.

Application of the following simple method (steps 1 and 2) allows us to easily calculate the charge, called the *formal charge,* on any atom in an organic structure. The method assumes that all non-bonded electrons are assigned to the nucleus they surround and that bonded electrons are equally shared between the bonded nuclei.

1. Add *all* the non-bonded electrons and *half* of the bonded electrons at the atom.
2. Subtract the sum from the group number of the element in the periodic table. The difference equals the formal charge on the atom.

What this calculation does is to compare the number of valence electrons assigned to a given covalently bonded atom with the number of valence electrons in the neutral (uncharged) atom. If the numbers are the same, then the covalently bonded atom is also neutral and the formal charge is 0. If there is one more valence electron assigned to the covalently bonded atom, then the charge at that site is -1; if one less, the charge at that site is $+1$. (Be sure to note that the *total* number of electrons around a given atom is *always* 8 or 2 in compliance with the octet rule.) Routinely calculating or verifying the formal charges in organic structures is a good habit to get into and will make mastery of organic chemistry much easier.

Example:

Calculate the formal charge on the nitrogen atom of *ammonia* and the *ammonium ion:*

Ammonia

$$2 \text{ non-bonded e}^- + \frac{6 \text{ bonded e}^-}{2} = 5$$

Group # of nitrogen: 5

Formal charge = 5 - 5 = 0

Ammonium ion

$$0 \text{ non-bonded e}^- + \frac{8 \text{ bonded e}^-}{2} = 4$$

Group # of nitrogen: 5

Formal charge = 5 - 4 = 1

Example:

Calculate the formal charge on either carbon atom of ethylene, C_2H_4:

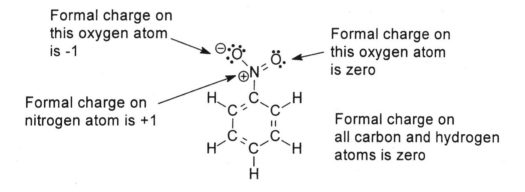

0 non-bonded e⁻ + $\dfrac{\text{8 bonded e}^-}{2}$ = 4

Group # of carbon: 4

$\boxed{\text{Formal charge = 4 - 4 = 0}}$

Example:

Calculate the formal charge on the indicated atoms of nitrobenzene, $C_6H_5NO_2$:

Formal charge on this oxygen atom is -1

Formal charge on this oxygen atom is zero

Formal charge on nitrogen atom is +1

Formal charge on all carbon and hydrogen atoms is zero

Example:

Calculate the formal charges on the atoms of the nitrate ion, NO_3^-:

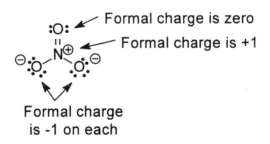

Formal charge is zero

Formal charge is +1

Formal charge is -1 on each

Chapter 8
Molecular Shapes: Valence Shell Electron Pair Repulsion Theory

The shape of covalent molecules can be reliably predicted by applying the *Valence Shell Electron Pair Repulsion* theory, often referred to by its acronym, *VSEPR*. This elegant and successful theory is based on the simple notion that valence shell electron pairs will assume angles as far as possible from each other to minimize the destabilizing effect of like-charge repulsion between these electron pairs. The shape of the molecule is determined by the application of this principle at the central atom. For example, the methane molecule has four bonded electron pairs in the valence shell of the central carbon atom. A tetrahedral arrangement of these bonded electron pairs gets them as far from each other as possible, thus determining the tetrahedral shape of methane and all other saturated hydrocarbons. (A tetrahedron is a pyramid having equilateral triangles for all sides. If the central carbon atom is situated exactly in the center of this geometric shape, the bonds to the four hydrogen atoms of methane are directed to the corners of the tetrahedron, with bond angles of 109.5°.)

Methane
CH_4

All bond angles: 109.5°
Molecular Shape: tetrahedral

In the water molecule, the central oxygen atom has two non-bonded electron pairs and a bonded electron pair to each of the two H atoms, for a total of four electron pairs on the central atom. As for methane above, this total of four pairs also assumes roughly tetrahedral angles to maximize separation and minimize repulsion. Note that since the non-bonded pairs occupy a little more room, and cause more repulsion, than the bonded pairs, the H-O-H angle is in fact, about 105°, a little less than the perfect tetrahedral angle of 109.5°. (Note also that the *shape* of the water molecule is correctly described as *bent planar*, since the molecular shape is determined by the positions of the three *atomic nuclei* which comprise this molecule; three points can only define a planar shape.)

Water
H_2O

Bond angle: 105°
Molecular Shape: bent planar

Carbon dioxide, CO_2, is linear. A bond angle of 180° maximizes the separation of the two electron pairs on either side of the central carbon atom. The bonds of the acetylene molecule are also linear with 180° angles, for the same reason. (This bond angle occurs in all molecules, called *alkynes*, which have a carbon-to-carbon triple bond.)

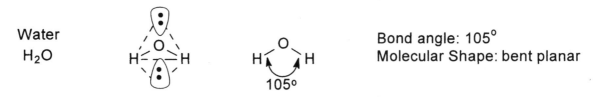

Carbon dioxide
CO_2

Molecular shape: linear

Acetylene
C_2H_2

Molecular shape: linear

Likewise, the ethylene molecule can minimize electron pair charge repulsion by assuming a planar shape with 120° angles between all bonded atoms. Organic compounds having carbon-to-carbon double bonds are called *alkenes*. Bond angles of approximately 120° are common to all alkenes in compliance with the VSEPR theory.

Ethylene
C_2H_4

All angles approximately 120°

Molecular shape: planar

The VSEPR theory does an excellent job of correctly predicting the shape of both organic and inorganic covalently bonded molecules and complex ions. The method is applied by following these simple steps:

1. Identify the central atom.
2. Count the number of *atoms* directly bound to the central atom.
3. Count the number of *non-bonded* electron pairs on the central atom.
4. Add 2. and 3.
5. Use the table below to determine the geometric arrangement of bonded atoms and non-bonded pairs. (This is the geometry and angle of maximum electron pair separation.)
6. Report the molecular shape by considering *only* the geometric arrangement of the *central atom and atoms directly bonded to the central atom* derived from step 5.

The following table summarizes the maximum separation geometry of electron pairs on the central atom. Here A stands for the central atom, and X for the sum of non-bonded electron pairs on the central atom and atoms directly bonded to it. The shape of the *molecule* is then derived from this table by considering only the positions of the central atom and atoms directly bonded to it.

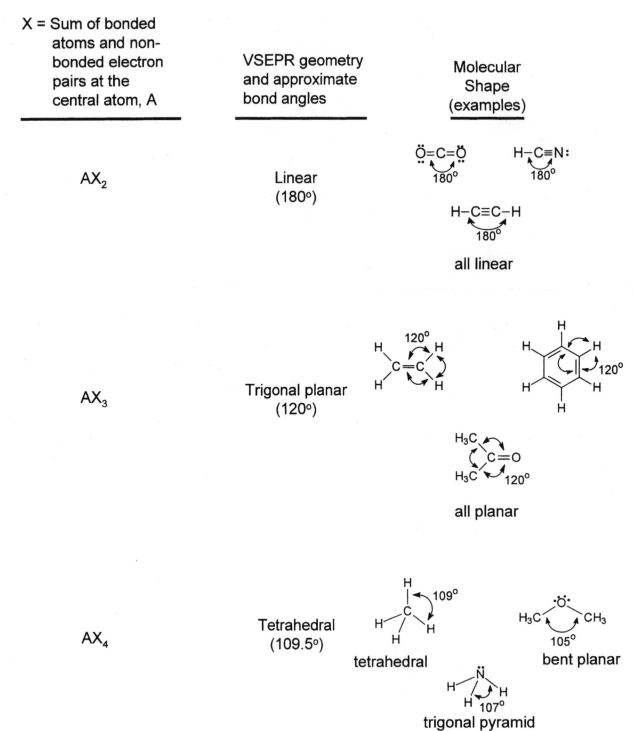

X = Sum of bonded atoms and non-bonded electron pairs at the central atom, A	VSEPR geometry and approximate bond angles	Molecular Shape (examples)
AX_2	Linear (180°)	all linear
AX_3	Trigonal planar (120°)	all planar
AX_4	Tetrahedral (109.5°)	tetrahedral / bent planar / trigonal pyramid

Fig 8-1: Molecular geometry and the Valence Pair Electron Pair Repulsion theory.

Chapter 9
Atomic Orbitals

The interaction of an electron with a proton to produce a hydrogen atom also produces a system of *orbitals* surrounding the nucleus. These orbitals are regions of specific geometry surrounding the proton (nucleus), which electrons, if present, will occupy. (More accurately, the orbitals are regions where the *probability* of finding an electron is high. This is so because modern physics tells us that the exact location of an electron cannot be pinpointed, but only described as a mathematical probability.)

The various orbitals generated in this way are designated by lower case letters *s,p,d* and *f*. These each have different shapes and symmetries about the central nucleus. The *s* orbitals are spherically symmetric about the nucleus, while *p* orbitals are dumbbell shaped, mutually perpendicular (orthogonal) and centered at the nucleus. The shapes of the *d* and *f* orbitals are more complicated but these orbitals will not concern us at this time. (The letter designations are terms describing spectral lines from experimental physics and stand for *s*harp, *p*rincipal, *d*iffuse and *f*undamental).

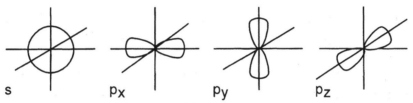

Fig 9-1: Shapes of s and p orbitals.

These orbitals become filled with electrons in an orderly sequence as the atomic number of the element, and therefore the number of electrons, increases. For the first 18 elements in the periodic table, the order of filling is 1s, 2s, 2p, 3s, and 3p. Here the numbers 1,2, and 3 refer to the electron shells which are filled in sequential order, at least for these light elements. In general, orbitals are filled when they contain two electrons. The s orbitals are filled to capacity with 2 electrons, while the mutually perpendicular p orbitals are actually composed of p_x, p_y, and p_z orbitals and therefore require a total of 6 electrons (two in each of the three p orbitals) to be filled. The order of filling is more completely indicated by the sequence shown: $1s^1$, $1s^2$, $2s^1$, $2s^2$, $2p^1_x$, $2p^1_y$, $2p^1_z$, $2p^2_x$, $2p^2_y$, $2p^2_z$, $3s^1$, $3s^2$, $3p^1_x$, $3p^1_y$, $3p^1_z$, $3p^2_x$, $3p^2_y$, $3p^2_z$, where the number of electrons in a given orbital is indicated by the superscript. (The sequence of filling the p orbitals need not concern us here.)

21

In another common presentation, the sequence of orbital filling is indicated by the arrow diagrams seen below. Here the electrons are represented by up or down arrows and the orbitals by horizontal lines. The order of filling for the first 10 elements is shown below:

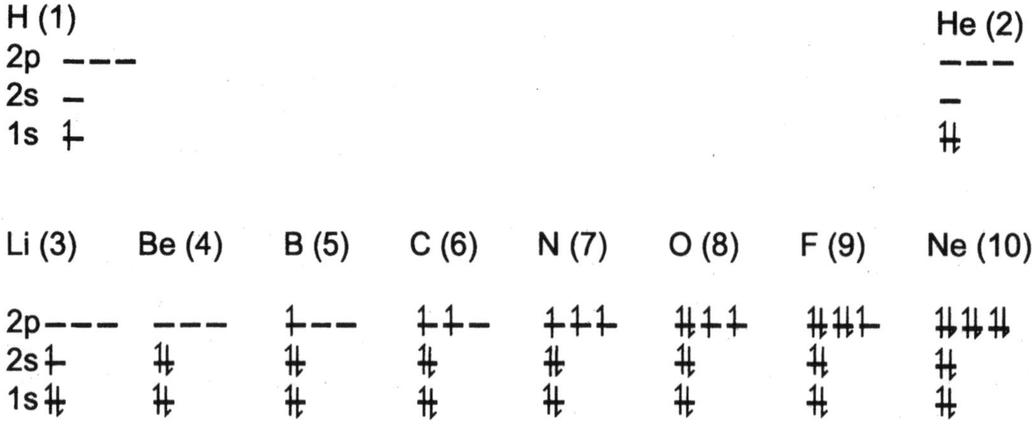

H (1)   He (2)
2p ─ ─ ─   ─ ─ ─
2s ─   ─
1s ⥮   ⥮

Li (3)   Be (4)   B (5)   C (6)   N (7)   O (8)   F (9)   Ne (10)

2p ─ ─ ─   ─ ─ ─   ⥮ ─ ─   ⥮ ⥮ ─   ⥮ ⥮ ⥮   ⥮ ⥮ ⥮   ⥮ ⥮ ⥮   ⥮ ⥮ ⥮
2s ⥮   ⥮   ⥮   ⥮   ⥮   ⥮   ⥮   ⥮
1s ⥮   ⥮   ⥮   ⥮   ⥮   ⥮   ⥮   ⥮

Fig 9-2: Electron configurations for the first 10 elements.

The first shell is filled by two electrons occupying the 1s orbital, to produce helium. The second shell is filled by the two 2s electrons plus the six 2p electrons, for a total of 8, to produce the noble gas neon. Thus, this quantum mechanical view of atomic structure is reconciled with the octet rule: *Filled electron shells can be described by filled orbitals of which the shells are composed.* The *ionic* chemistry of the light elements can be well understood based on gain or loss of electrons to or from partially filled orbitals to achieve the electron configuration of helium or neon or the heavier noble gases. This is shown in Figure 9-3 which describes the transfer of one electron from a lithium atom to a fluorine atom to form lithium fluoride:

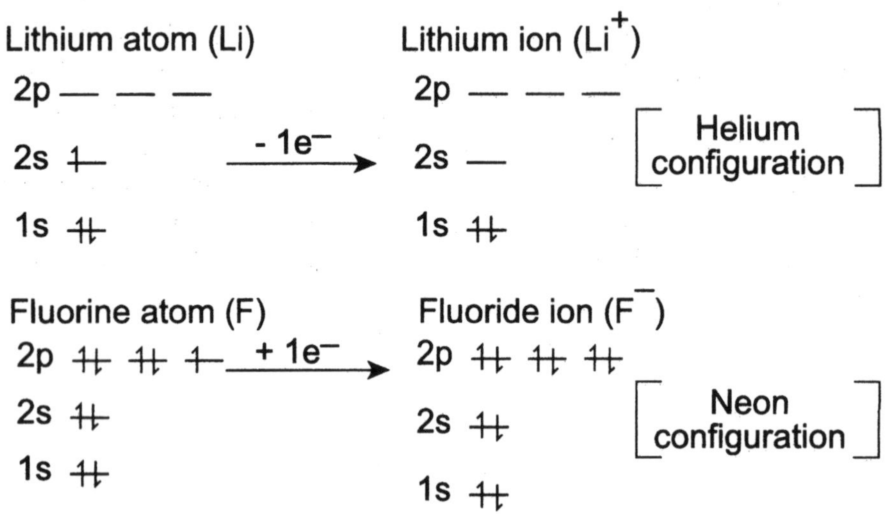

Lithium atom (Li)   Lithium ion (Li$^+$)
2p ─ ─ ─   2p ─ ─ ─
                                                      ⎡ Helium   ⎤
2s ⥮ $- 1e^-$ →   2s ─   ⎣ configuration ⎦

1s ⥮⥮   1s ⥮⥮

Fluorine atom (F)   Fluoride ion (F$^-$)
2p ⥮⥮ ⥮⥮ ⥮ $+ 1e^-$ →   2p ⥮⥮ ⥮⥮ ⥮⥮
                                                      ⎡ Neon   ⎤
2s ⥮⥮   2s ⥮⥮   ⎣ configuration ⎦

1s ⥮⥮   1s ⥮⥮

Fig 9-3: Electron transfer from lithium to fluorine to produce lithium fluoride.

Chapter 10
Orbital Hybridization

The *ion-forming* chemical reactivity of atoms can be well understood based on the atomic orbital model, but *covalent* bonding cannot be so easily explained. Carbon, for example, forms the simple molecule methane, CH_4, where one carbon atom reacts with four hydrogen atoms to produce a molecule having four *equivalent* covalent bonds. It is difficult to explain this result using the atomic orbital diagram for carbon, seen above in Figure 9-2, which shows the valence shell having a filled 2s orbital and two p orbitals having a single electron each. Direct overlap of these with an s orbital from each of four hydrogen atoms would hardly produce four equivalent bonds.

To explain the covalent bonding of carbon and other elements, the concept of *orbital hybridization* was proposed. Here, it is theorized that as a carbon atom reacts to form a saturated covalent molecule, such as methane, there occurs a mixing, or hybridization, of the one 2s and three 2p orbitals to produce 4 *equivalent* hybridized orbitals. These hybrid orbitals each have one part s and three parts p character, and are called *sp³* for that reason. These four equivalent hybrid orbitals permit a rational model for the four equivalent bonds of methane. Each of these four sp³ orbitals would carry one of the four valence electrons of carbon. Overlap of these with four hydrogen atoms results in the methane molecule. This sp³ hybridization is common to all *saturated* carbon compounds,—those in which carbon is bound to the maximum of four other atoms. These saturated compounds are also called *alkanes,* in which all bonds result from orbital overlap along the axis of a line joining the bonded nuclei. These bonds are called *sigma* (σ) bonds. Tetrahedral bond angles of 109.5° characterize the sigma bonds of the alkanes, consistent with the Valence Shell Electron Pair Repulsion theory (VSEPR).

In the diagram below representing orbital hybridization, atomic and hybrid orbitals are represented as *squares,* and electrons as *arrows* pointing up or down. (The direction of the arrow refers to electron *spin.* A covalent bond is formed from a pairing of electrons having opposite spins.)

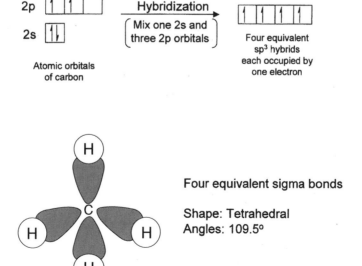

Fig 10-1: Orbital hybridization of carbon in methane and all alkanes.

Carbon atoms can also form unsaturated compounds, called *alkenes,* in which a carbon atom is bonded to three other atoms by two single bonds and one double bond. Bonding in such compounds is explained based on hybridization of the 2s orbital with only two of the 2p orbitals. This produces three sp² hybrid orbitals, each having one-part s and two- parts p character, with one leftover (un-hybridized) p-orbital. Overlap of these hybrid orbitals with orbitals from bonded atoms results in three sigma bonds, while overlap of the dumbbell shaped p-orbital with the p-orbital of an adjacent carbon atom, forms a *pi* (π) bond. This pi bond results from overlap of the p-orbital lobes above and below the axis of the sigma bond. Bonding in the simplest member of the alkenes, called ethylene, is shown below. The bonded atoms are separated by an angle of approximately 120° in ethylene and all other alkenes, as predicted by the VSEPR theory.

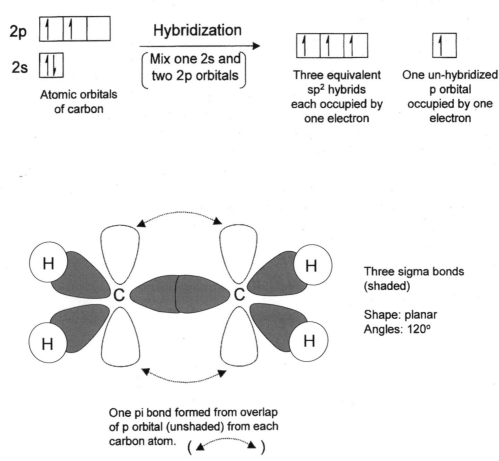

Fig 10-2: Orbital hybridization of carbon in ethylene and all alkenes.

Carbon atoms can also bond to other carbons by triple bonds. This produces compounds called *alkynes.* The simplest member of this class of compounds is called acetylene (or ethyne), H—C≡C—H. Here, mixing of the 2s orbital with only one of the 2p orbitals creates two sp hybrids required for the two sigma bonds. This leaves two un-hybridized p-orbitals left over, which overlap with p-orbitals on an adjacent carbon atom forming two perpendicular (orthogonal) pi bonds. The bond angles in alkynes are 180°, in accord with the VSEPR theory, for acetylene and all other alkynes.

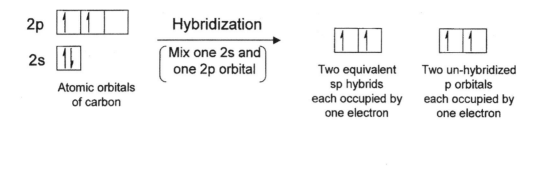

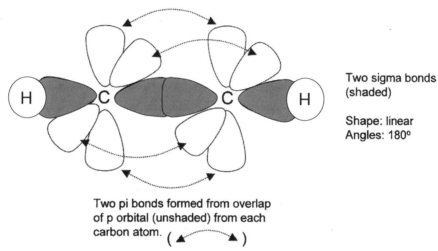

Fig 10-3: Orbital hybridization of carbon in ethyne and all alkynes.

This hybridization scheme cannot be done willy-nilly and must obey certain rules. In each case the total number of hybrid orbitals is the same as the total number of atomic orbitals combined to form them, and the sum of the hybridized and unhybridized orbitals is always equal to the original number of atomic orbitals. Thus, the total number of orbitals is *conserved*. In the preceding diagrams, the number of orbitals, represented as squares, is always the same before and after hybridization. (The number of electrons, represented by arrows, before and after hybridization must be unchanged as well, to comply with conservation of mass and charge.)

The simple table below summarizes the shapes of molecules based on the valence shell electron pair repulsion theory, VSEPR, and orbital hybridization schemes for several important classes of hydrocarbons.

COMPOUND CLASS	VSEPR TYPE	HYBRIDIZATION	SHAPE (approximate bond angle)
Alkanes	AX_4	sp^3	Tetrahedral (109.5°)
Alkenes	AX_3	sp^2	Planar (120°)
Alkynes	AX_2	sp	Linear (180°)

Fig 10-4: Hybridization and bond angles for classes of hydrocarbons.

As we will see in the following chapters, only a few simple molecular shapes and hybridization schemes occur in *all* organic chemical structures, and not just in hydrocarbons. *Bond angles and molecular shapes can be described either by electron pair repulsion arguments or the orbital hybridization of their sigma bonds.*

Chapter 11
The Functional Groups

Organic compounds are composed of covalently bonded atoms. Generally, chains of linked carbon atoms with associated hydrogen atoms form the backbone of an organic molecule. These occur in various sizes and shapes, including straight and branched chains and cyclic structures of various ring sizes. More complex structures can include several or all of these variations in the same molecule.

Certain common groupings of atoms are often appended to this hydrocarbon backbone. These atomic groupings, called *functional groups,* tend to occur repetitively in a wide variety of organic structures. There are only about a dozen very common functional groups and perhaps another dozen which are less common. The chemical and physical properties of an organic compound are determined in large measure by the presence of one or more of these functional groups in the molecular structure. Knowledge of the type, number, and proximity of functional groups in a given structure often allows a chemist to make accurate predictions about the properties of that compound. These include the chemical reactions that compound will undergo, along with estimates of the rate of reactivity, and physical properties, such as solubility, melting and boiling temperatures, color, and even odor.

Each of the various functional groups undergoes less than a dozen different reactions which utilize one or more specific reagents and a general set of conditions to convert that group into another functional group. These conversions often occur (to a first approximation) independently of the structure of the rest of the molecule. *Mastery of a large bulk of organic chemical knowledge consists of learning the important functional groups and the major reactions of each.* Since these reactions are characteristic of a given group, they are often referred to as *type reactions.* The next few pages will introduce the names, structure, and shape of the most common functional groups, and a few important chemical reactions and physical properties of each. This section also introduces the symbol **R**, which represents a *generalized* carbon-containing group. The symbol is frequently used in organic chemistry to emphasize a *general* property or chemical reactivity. Chapters 18-32 in part II of this book will investigate the chemistry and physical properties of the functional groups in greater detail.

1. The Hydrocarbons (Alkanes, Alkenes, Alkynes and Aromatics)

These compounds are composed only of carbon and hydrogen atoms. Each type is presented in greater detail in later chapters.

The Paraffins:(Alkanes)

The *paraffin* hydrocarbons are characterized by *single* bonded carbon atoms. They have a tetrahedral shape and sp³ hybridization. They are also called *alkanes* and *saturated hydrocarbons,* the latter term referring to the fact that each carbon atom is saturated with four (the maximum number) of bound atoms. The chemical reactivity of alkanes is low; they undergo relatively few chemical reactions. They do, however, react with oxygen to produce CO_2 and H_2O, and undergo substitution reactions with halogens like Cl_2. These alkane hydrocarbons often form the structural backbone of other more reactive and complex compounds.

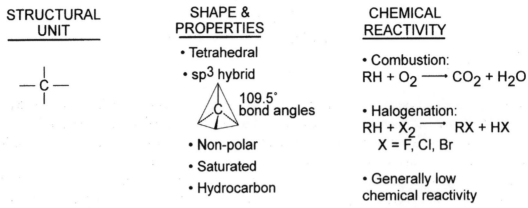

STRUCTURAL UNIT	SHAPE & PROPERTIES	CHEMICAL REACTIVITY
$-\overset{\vert}{\underset{\vert}{C}}-$	• Tetrahedral • sp³ hybrid 109.5° bond angles • Non-polar • Saturated • Hydrocarbon	• Combustion: $RH + O_2 \longrightarrow CO_2 + H_2O$ • Halogenation: $RH + X_2 \longrightarrow RX + HX$ $X = F, Cl, Br$ • Generally low chemical reactivity

Fig 11-1: The alkanes.

The Olefins (Alkenes)

These compounds have a double bond linking a pair of adjacent carbon atoms. The carbon-carbon sigma bonds are sp² hybridized and are planar with 120° bond angles. The pi bond results from the overlap of p-orbitals on the bonded carbons. These *olefins* or *alkenes* are *unsaturated* since the carbon atoms are bonded to *less* than the maximum number of other atoms. They are characterized by addition and oxidation reactions, in which atoms and atomic groupings add across the pi bond, and oxidative cleavage reactions in which the molecule is oxidized (by accepting one or more oxygen atoms) and then cleaved into two fragments at the double bond.

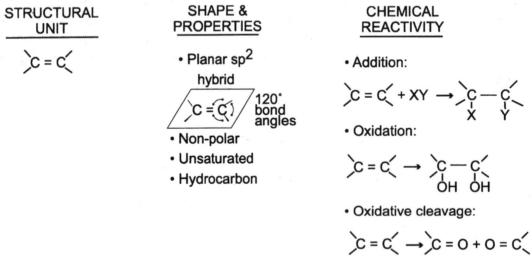

STRUCTURAL UNIT	SHAPE & PROPERTIES	CHEMICAL REACTIVITY
$\overset{\diagdown}{\underset{\diagup}{C}} = \overset{\diagdown}{\underset{\diagup}{C}}$	• Planar sp² hybrid 120° bond angles • Non-polar • Unsaturated • Hydrocarbon	• Addition: • Oxidation: • Oxidative cleavage:

Fig 11-2: The alkenes.

The Acetylenes (Alkynes)

The *acetylenes* or *alkynes* are highly unsaturated, possessing a carbon-to-carbon *triple* bond, composed of two pi bonds and an sp hybridized sigma bond joining the bonded atoms along a straight line, at a 180° bond angle.

As many as four atoms or groups can add across the two pi bonds of the *alkynes* to produce a saturated product. Reacting much like alkenes, the alkynes undergo both addition and oxidation reactions, including oxidative cleavage.

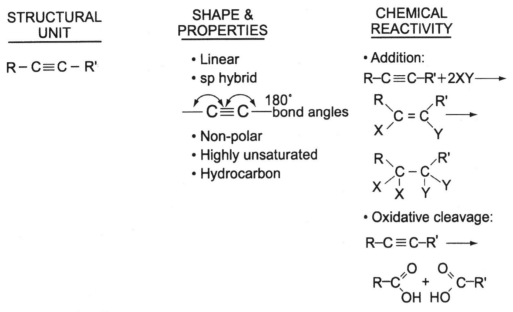

Fig 11-3: The alkynes.

2. Aromatic Compounds

The *aromatic* hydrocarbons have an alternating system of single and double bonds completely around a closed ring system. (Certain other criteria must also be met for a compound to be classified as aromatic.) This bond arrangement results in a special stabilization for aromatic molecules. Consequently, aromatic compounds undergo different chemical reactions from non-aromatic, or *aliphatic,* compounds. Aromatic compounds generally react to substitute various atoms or atomic groupings for ring hydrogen atoms. These substitution reactions preserve the stable character of the aromatic ring system.

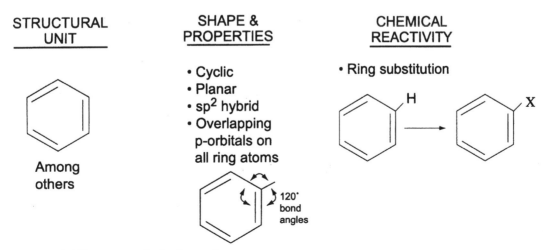

Fig 11-4: The aromatic hydrocarbons.

3. The Alkyl Halides

Alkyl halides have one or more *halogen* atoms substituted for the hydrogen atoms of an alkane hydrocarbon. *Alkyl* means derived from an alkane. These compounds undergo *substitution* reactions in which the halide is lost and replaced by a variety of other atoms or groups; they also undergo elimination reactions—losing the elements of a halogen acid (HX) from adjacent carbon atoms. In addition, alkyl halides react with magnesium metal and other active metals to produce *Grignard* reagents which react as *carbanions*. (Carbanions are reactive intermediates in which a carbon atom carries a negative charge.) The reactions of Grignard reagents form one of the most important classes of organic reactions.

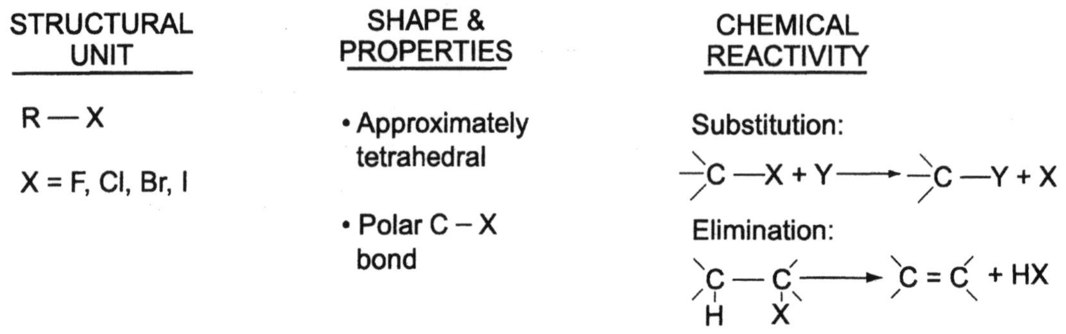

STRUCTURAL UNIT	SHAPE & PROPERTIES	CHEMICAL REACTIVITY
R — X X = F, Cl, Br, I	• Approximately tetrahedral • Polar C – X bond	Substitution: $-\overset{\|}{\underset{\|}{C}}-X + Y \longrightarrow -\overset{\|}{\underset{\|}{C}}-Y + X$ Elimination: $\underset{H}{\overset{\|}{C}} - \underset{X}{\overset{\|}{C}} \longrightarrow C = C + HX$

Fig 11-5: The alkyl halides.

4. The Alcohols

The *alcohols* possess the —OH group, directly linked to a carbon atom. They have the general formula, ROH, and are thus organic analogs of water, HOH. Like water, they are polar and engage in hydrogen bonding. Alcohols are highly reactive and participate in many reactions including oxidation, substitution, and dehydration (loss of water).

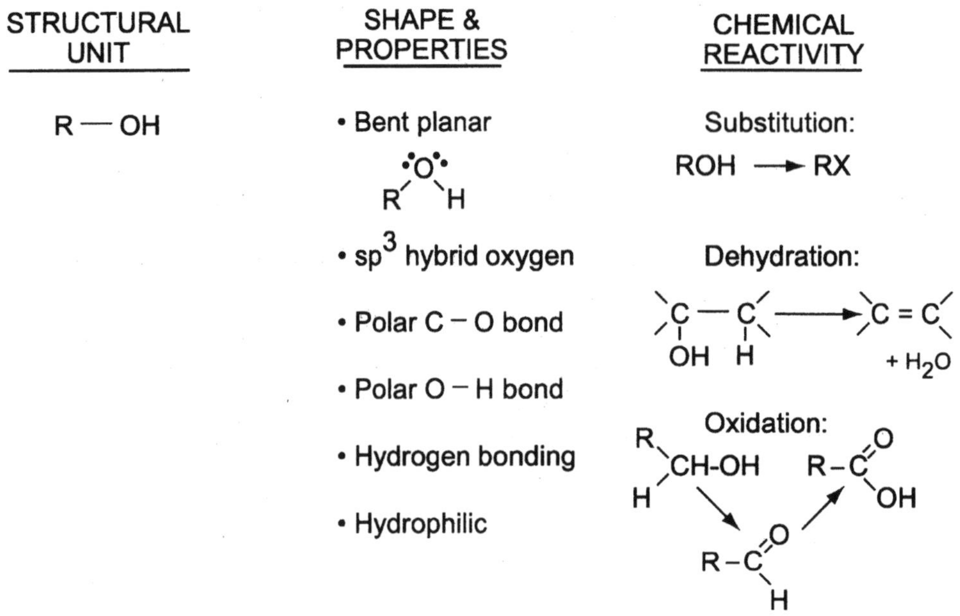

STRUCTURAL UNIT	SHAPE & PROPERTIES	CHEMICAL REACTIVITY
R — OH	• Bent planar $\overset{..}{\underset{R \quad H}{O}}$ • sp^3 hybrid oxygen • Polar C – O bond • Polar O – H bond • Hydrogen bonding • Hydrophilic	Substitution: ROH ⟶ RX Dehydration: $\underset{OH \quad H}{\overset{\|}{C} - \overset{\|}{C}} \longrightarrow C = C \;\; + H_2O$ Oxidation: $\underset{H}{\overset{R}{>}}CH\text{-}OH \qquad R\text{-}C\overset{\nearrow O}{\searrow OH}$ $R\text{-}C\overset{\nearrow O}{\underset{H}{}}$

Fig 11-6: The alcohols.

5. The Ethers

Ethers have the general formula R—O—R′ and are characterized by the presence of a C—O—C bond. In general, they are relatively un-reactive, participating in only a few chemical reactions. They can undergo substitution reactions, but only under conditions of very high temperature and acid strength. Three-membered ring ethers, called epoxides, are an exception, however. They readily engage in reactions which result in opening the strained ring of these compounds.

STRUCTURAL UNIT	SHAPE & PROPERTIES	CHEMICAL REACTIVITY
ROR′	• Bent planar 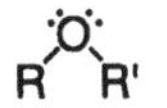 $\ddot{\underset{R \quad R'}{O}}$ • sp³ hybrid oxygen • Polar covalent C-O bond	• Low reactivity • Cleavage can occur with difficulty • Three-membered ring cyclic ethers (epoxides) are reactive 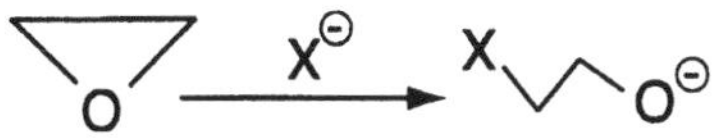

Fig 11-7: The ethers.

6. Carbonyl-Containing Functional Groups

The *carbonyl group* has a carbon atom double bonded to an oxygen atom. This occurs in a variety of organic functional groups, including aldehydes, ketones, carboxylic acids, esters, and amides, among others. (Each of these will be discussed and described further in various chapters in Part 2 of this book.) The properties and chemical reactivity of all these functional groups is due in large part to the polar, unsaturated carbonyl group.

Aldehydes

Aldehydes have the general formula RCHO. They undergo many substitution and addition reactions, as well as reduction, reductive alkylation and oxidation. (In this context, reduction refers to accepting more *hydrogen* atoms; oxidation to accepting more *oxygen* atoms.)

31

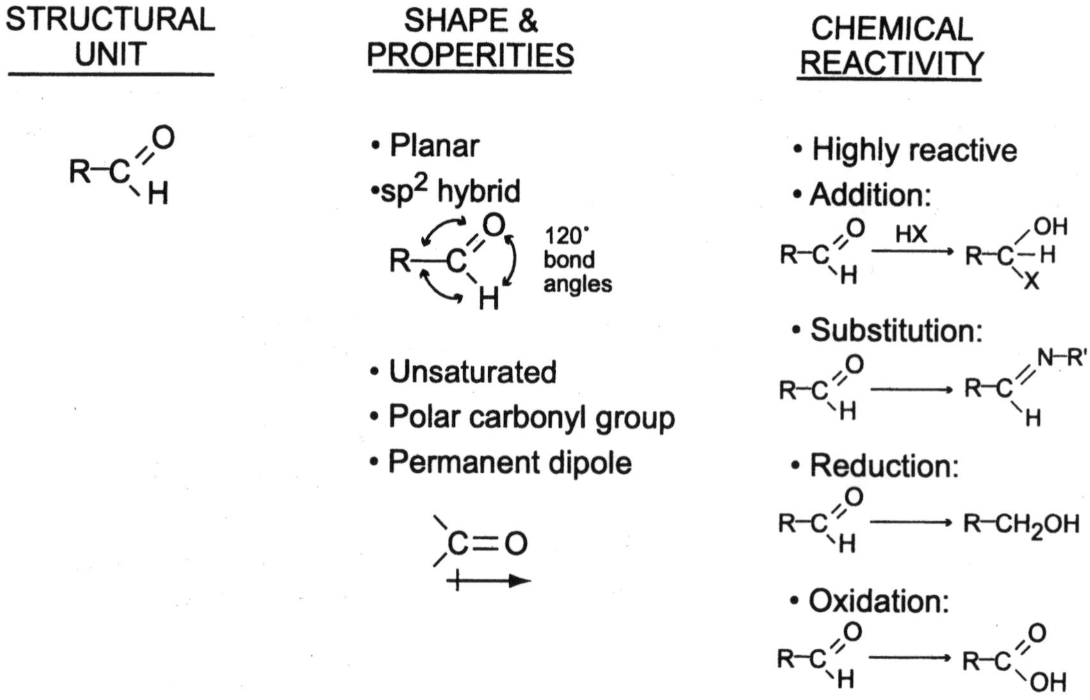

STRUCTURAL UNIT	SHAPE & PROPERITIES	CHEMICAL REACTIVITY

$$R-C\overset{O}{\underset{H}{\lesssim}}$$

- Planar
- sp^2 hybrid

$$R-C\overset{O}{\underset{H}{\lesssim}} \quad 120° \text{ bond angles}$$

- Unsaturated
- Polar carbonyl group
- Permanent dipole

$$\overset{\diagdown}{\underset{\diagup}{C}}=O$$

- Highly reactive
- Addition:

$$R-C\overset{O}{\underset{H}{\lesssim}} \xrightarrow{HX} R-C\overset{OH}{\underset{X}{\overset{|}{-}}}H$$

- Substitution:

$$R-C\overset{O}{\underset{H}{\lesssim}} \longrightarrow R-C\overset{N-R'}{\underset{H}{\diagup}}$$

- Reduction:

$$R-C\overset{O}{\underset{H}{\lesssim}} \longrightarrow R-CH_2OH$$

- Oxidation:

$$R-C\overset{O}{\underset{H}{\lesssim}} \longrightarrow R-C\overset{O}{\underset{OH}{\lesssim}}$$

Fig 11-8: The aldehydes.

Ketones

Ketones have the general formula RCOR′. They are similar to aldehydes in chemical structure and reactivity, but ketones generally do not undergo oxidation, and do not undergo substitution as easily as aldehydes.

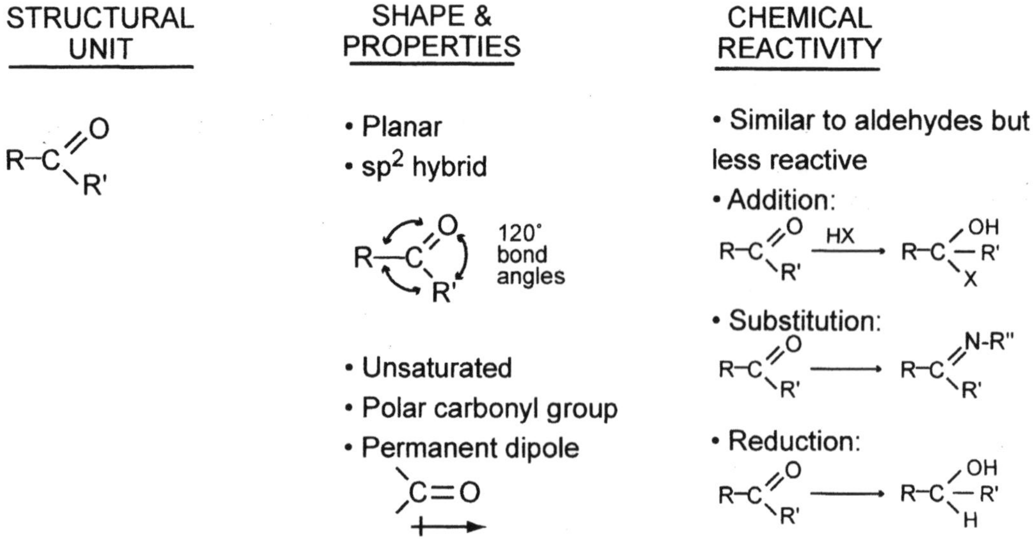

STRUCTURAL UNIT	SHAPE & PROPERTIES	CHEMICAL REACTIVITY

$$R-C\overset{O}{\underset{R'}{\diagup}}$$

- Planar
- sp^2 hybrid

$$R-C\overset{O}{\underset{R'}{\lesssim}} \quad 120° \text{ bond angles}$$

- Unsaturated
- Polar carbonyl group
- Permanent dipole

$$\overset{\diagdown}{\underset{\diagup}{C}}=O$$

- Similar to aldehydes but less reactive
- Addition:

$$R-C\overset{O}{\underset{R'}{\lesssim}} \xrightarrow{HX} R-C\overset{OH}{\underset{X}{\diagup}}R'$$

- Substitution:

$$R-C\overset{O}{\underset{R'}{\lesssim}} \longrightarrow R-C\overset{N-R''}{\underset{R'}{\diagup}}$$

- Reduction:

$$R-C\overset{O}{\underset{R'}{\lesssim}} \longrightarrow R-C\overset{OH}{\underset{H}{\diagup}}R'$$

Fig 11-9: The ketones.

Carboxylic Acids

Carboxylic acids are the acids of organic chemistry. They have the general formula RCOOH, and ionize to some extent, liberating a proton and conferring an acidic pH in water solution. They will ionize completely to form water and salts upon reaction with bases like sodium hydroxide or ammonia. They can be converted to aldehydes and alcohols by reducing agents, and also undergo substitution reactions to yield various *carboxylic acid derivatives* including esters, amides, acid halides, and anhydrides.

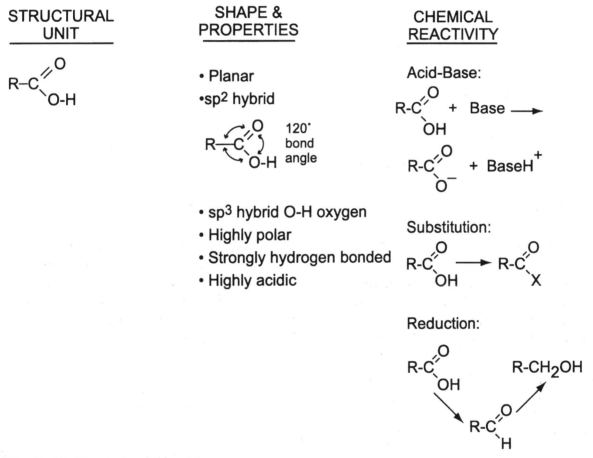

Fig 11-10: The carboxylic acids.

Carboxylic Acid Derivatives

These include *esters, amides, acid halides,* and *anhydrides.* They can be hydrolyzed (reacted with water) to produce carboxylic acids, and they can all be inter-converted by substitution reactions.

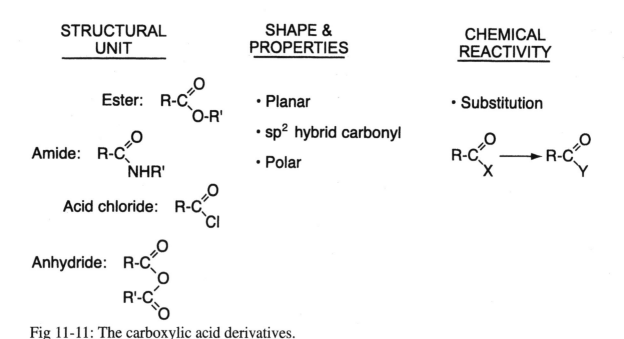

STRUCTURAL UNIT	SHAPE & PROPERTIES	CHEMICAL REACTIVITY

Ester: $R-C\begin{smallmatrix}O\\O-R'\end{smallmatrix}$

Amide: $R-C\begin{smallmatrix}O\\NHR'\end{smallmatrix}$

Acid chloride: $R-C\begin{smallmatrix}O\\Cl\end{smallmatrix}$

Anhydride: $R-C\begin{smallmatrix}O\\O\end{smallmatrix}$ $R'-C\begin{smallmatrix}O\end{smallmatrix}$

• Planar

• sp² hybrid carbonyl

• Polar

• Substitution

$R-C\begin{smallmatrix}O\\X\end{smallmatrix} \longrightarrow R-C\begin{smallmatrix}O\\Y\end{smallmatrix}$

Fig 11-11: The carboxylic acid derivatives.

7. Amines

Amines have the general structural formulas, RNH_2, $RNHR'$, and $RNR'R''$, for primary (1°), secondary(2°), and tertiary(3°) amines, respectively. They have at least one carbon atom directly attached to a nitrogen atom by a single bond. They are organic analogs of ammonia, NH_3, and are the bases of organic chemistry. Amines confer a basic pH when dissolved in water, since they react with water to produce an alkyl ammonium hydroxide. Amines also participate in substitution reactions, replacing various leaving groups, shown as X below, at both saturated and acyl (carbonyl) carbon.

STRUCTURAL UNIT	SHAPE & PROPERTIES	CHEMICAL REACTIVITY

$R-NH_2$ 1°
$RR'NH$ 2°
$RR'R''N$ 3°

• Trigonal pyramid

$$\underset{R'}{\overset{\ddot{N}}{R-\!\!\!-\!\!\!-R''}}$$

• Nitrogen sp³ hybrid
• Polar C-N and N-H bonds
• 1° and 2° amines hydrogen bond
• Highly basic

• Acid - Base:
$R-NH_2 + HX \longrightarrow RNH_3^{\oplus} X^{\ominus}$

• Nucleophilic substitution:
$R-NH_2 + -C-X \longrightarrow$

$R-NH-C- + HX$

Fig 11-12: The amines.

Chapter 12
Physical Properties and Molecular Structure

A high level of correlation exists between the structure of organic molecules and their physical properties, including boiling and melting temperatures, degree of solubility, viscosity, and even color. This is important since chemical substances present themselves to us primarily by these very properties. We describe a newly observed material first and foremost by its physical state: Is it a solid, liquid, or gas at ordinary temperatures? Is it volatile, viscous, waxy, oily, and so forth? Is it soluble in water, like antifreeze and sugar, or is it insoluble? Is it white or colored? That we can correlate and explain these easily observable properties with the sub-microscopic structure of individual molecules is a good indication of the success of our chemical theories. Indeed, a smooth change can be observed in many such properties with systematic variation of molecular structure. Here are a few important generalizations:

1. *Increasing the length of a carbon chain results in an increase in boiling temperature (boiling point).*
   Explanation: Boiling involves overcoming the attractive forces between molecules that cause them to associate with each other (stick together) in the liquid state. The larger the molecule, the more such associations can occur, requiring more energy and higher temperatures to produce the vapor (gaseous) state. These intermolecular associations are indicated by the dashed lines in the diagrams below.

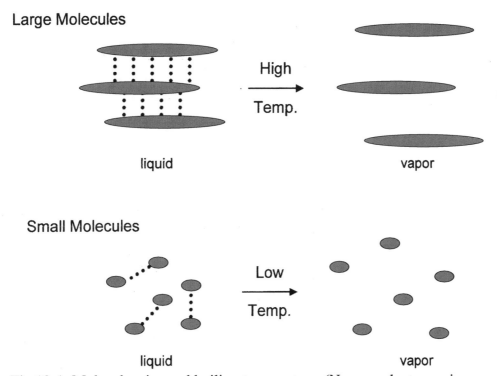

Fig 12-1: Molecular size and boiling temperature. (Non-covalent association shown by dotted lines)

2. *Polar molecules tend to have higher boiling points than comparably sized non-polar molecules. Polar molecules capable of hydrogen bonding have still higher boiling points.*

Explanation: Non-polar molecules can associate with each other only by relatively weak and transient forces called London or Van der Walls forces. The attraction between the opposite charged regions of molecules with permanent dipoles is stronger, and breaking them requires more thermal energy. Hydrogen-bonded molecules, like water and the alcohols, require still more energy to break these even stronger forces of inter-molecular association. The graph below shows the regular change in boiling temperature seen for homologous series of the normal alkanes (non-polar), aldehydes (polar), and alcohols (hydrogen-bonded), clearly demonstrating these trends.

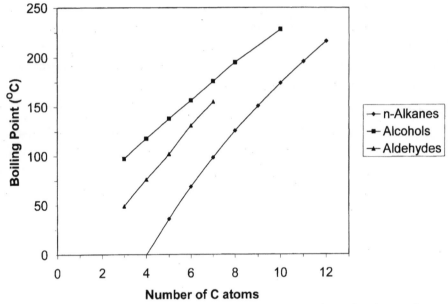

Fig 12-2: Boiling temperatures for homologous series of polar and non-polar compounds.

3. Viscosity increases with molecular size.

Explanation: Liquids tend to be viscous when molecules can stick together to form large aggregates. This aggregation occurs in large molecules in general and also in highly associated smaller molecules. Thus, high molecular weight hydrocarbons, like motor oil, and strongly hydrogen bonded substances, like antifreeze (ethylene glycol), tend to be viscous.

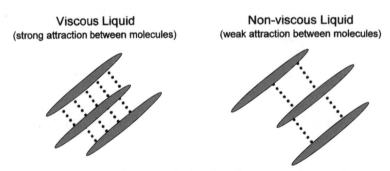

Fig 12-3: Molecular association in viscous and non-viscous liquids. (Non-covalent association shown by dotted lines.)

4. *Melting points tend to increase in a regular fashion with increasing molecular size within a homologous series. Often molecules that have rigid, symmetrical structures have exceptionally high melting points.*

Explanation: Melting involves the breakdown of a regular array of molecules in a solid crystalline lattice, to produce the randomly changing, transient associations characteristic of a liquid. In a solid crystal, neighboring molecules associate attractively at points of close approach. Large molecules have more points of association with neighbors than small molecules, and so melt at higher temperatures. Rigid and highly symmetrical molecules tend to fit together tightly into a regular crystal lattice, interacting strongly with their neighbors, and so tend to have especially high melting points.

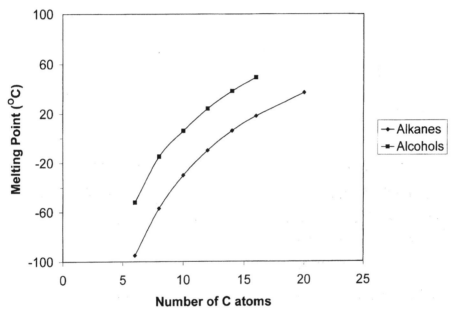

Fig 12-4a: Melting temperatures for homologous series of polar and non-polar compounds.

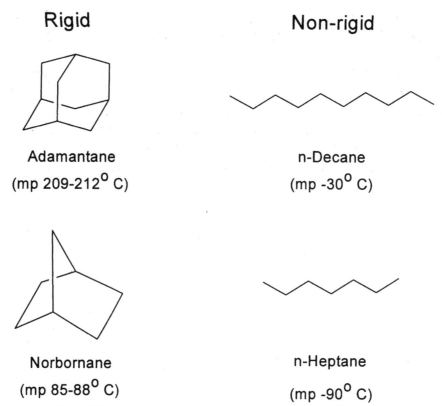

Rigid

Non-rigid

Adamantane

(mp 209-212° C)

n-Decane

(mp -30° C)

Norbornane

(mp 85-88° C)

n-Heptane

(mp -90° C)

Fig 12-4b: Melting temperatures of rigid and non-rigid compounds of the same carbon number.

5. *Molecules with a high ratio of the electronegative atoms N, O, and S to carbon atoms tend to be polar, and display the properties expected of polar molecules. These properties include high water solubility, high boiling points, and ability to undergo chemical reactions involving ionic mechanisms. Conversely, a low ratio of electronegative atoms to carbon atoms produces non-polar molecules which have the opposite properties: little or no water solubility, high solubility in hydrocarbon solvents, relatively low boiling points, and lack of chemical reactivity by ionic mechanisms. A regular change in many of these properties can be seen as the ratio of carbon atoms to electronegative atoms changes within a series of molecules.*

    Explanation: Bonds between carbon and the *heteroatoms,* N,S, and O, are polar due to electronegativity differences. Polar bonds often produce polar molecules. This causes these compounds to be water soluble, since polar solvents like water tend to dissolve polar solutes. This is often expressed by the statement "Like dissolves like." Polar bonds also result in relatively high boiling temperatures, due to the attraction between oppositely aligned dipoles on neighboring molecules. Chemical reactions proceeding through an ionic mechanism are favored by charge separation (polarity) in a reacting molecule, since a permanent partial charge is already present at the site of reaction. Conversely, molecules containing few electronegative atoms or polar bonds are non-polar, and the opposite properties are observed.

SOLUTE	SOLUBILITY (gr/100 ml H_2O)	RATIO(Carbon atoms/ Heteroatoms)
Alcohols:		
CH_3CH_2OH	∞	2:1
$CH_3CH_2CH_2CH_2OH$	8.3	4:1
$CH_3(CH_2)_4CH_2OH$	0.6	6:1
$CH_3(CH_2)_6CH_2OH$	0.05	8:1
Amines:		
$CH_3CH_2NH_2$	very soluble	2:1
$CH_3CH_2CH_2CH_2\text{-}NH_2$	very soluble	4:1
⬡–NH_2 (cyclohexylamine)	slightly soluble	6:1
⬡–$\overset{H}{N}$–⬡ (diphenylamine)	insoluble	12:1
Carboxylic acids:		
CH_3COOH	∞	2:2
$CH_3(CH_2)_2COOH$	∞	4:2
$CH_3(CH_2)_4COOH$	1.08	6:2
$CH_3(CH_2)_6COOH$	0.07	8:2
$CH_3(CH_2)_8COOH$	0.015	10:2

Fig 12-5: Water solubility for a series of alcohols, amines, and carboxylic acids.

6. *Odor and color are related to the presence of certain functional groups and other structural features of organic compounds.*

Aldehydes often have spicy and perfume odors, and are produced by many flowers to attract insects for pollination. Mercaptans, also called thiols, which contain the -SH group, all have a nauseating skunk-like odor, and one member of this group is indeed produced by skunks for its unique form of defense. Many low molecular weight carboxylic acids have a foul odor, including butyric acid which gives rancid butter its odor. Certain aliphatic amines are responsible for the characteristic odor of fish oils. Of course the strength of the odor falls off as the size of the molecule increases since large molecules tend to have a lower vapor concentration, and require a higher temperature to be detected by smell. It is, after all, the gaseous vapor that reaches our nose.

Explanation: Unlike certain other properties, explanations for specific odors are not simple, since the mechanism of smell involves interaction with a biochemical receptor and transmittal of a signal to the brain

molecules possessing the same functional group have a similar odor. This implies a specific bonding association of the functional group at the receptor site. An experienced organic chemist can often accurately infer the presence of certain functional groups of a compound based on smell.

Color results from the absorption of visible light by a chemical compound. The removal of a given color from white light produces a visual image of the complimentary color. For example, absorption (removal) of red light produces the appearance of a blue color, while the absorption of blue light causes a compound to be red. Most organic chemical compounds are white or colorless since they absorb only ultraviolet light, which we cannot see, rather than visible light. As we shall see in Chapter 34, compounds whose molecular structure has an extensive system of alternating single and multiple bonds absorb visible light, and are colored for that reason. Virtually all dyes possess that feature.

Alkane   Alkene   Alkyne   Al Cohen

Chapter 13
Lewis Acids and Bases

The reaction of a proton, H^+ with a hydroxide ion, OH^-, forms water:

$$H^{\oplus} + OH^{\ominus} \longrightarrow H_2O$$

The proton source in this reaction is called a *Bronsted acid,* while the hydroxide source is called a *Bronsted base.* A proton donor, in general, is defined as a Bronsted acid; a proton acceptor, as a Bronsted base.

In this reaction the proton is also an electron pair acceptor and the hydroxide is the electron pair donor. The new covalent bond of the water molecule is formed from the electron pair donated by the hydroxide ion.

Around the turn of the 20th century, the concept of acids and bases was expanded by Gilbert N. Lewis. He proposed that, *in general, an acid is defined as an electron pair acceptor while a base is defined as an electron pair donor.* This is one of the simplest but most important ideas in all of chemistry. It allowed a great many chemical reactions, including those not involving protons at all, to be included in the general category of acid-base reactions and thus simplified and unified chemistry. Many bond-forming reactions in organic chemistry involve the combination of an electron deficient species (the acid), often having a partial or full positive charge, with an electron rich species (the base), which can donate an electron pair to form a two-electron covalent bond. These reactants are referred to as *Lewis Acids* and *Lewis Bases,* respectively.

In this way most bond-forming events in organic chemistry can be explained by the *mutual attraction of opposite charges,* with a *negative* electron pair linking to an electron deficient, usually *positive,* center. Most of the reactions in this book can be understood, and accommodated within, this simple principle. The Lewis theory of acids and bases is one of the most important underlying concepts in chemistry.

A Lewis acid is often called an *electrophile* (having an affinity for negatively charged electron pairs) while a Lewis base is often referred to as a *nucleophile* (having an affinity for the positively charged nucleus). Note that common Lewis acids can have a full positive charge:

$$e.g. \quad H^{\oplus}, \quad CH_3^{\oplus}, \quad Br^{\oplus} \quad \text{(positive ions)}$$

or a partial positive charge:

e.g. $-\overset{|}{\underset{\delta+}{C}} - \underset{\delta-}{Br}$, $\overset{\diagdown}{\underset{\delta+}{C}} = \underset{\delta-}{O}$ (polar bonds)

acidic center

or be uncharged:

e.g.   Cl   Cl   F   F   (Only six electrons
      ＼ ／   ＼ ／   around central atom)
       Al   ,   B
       | |
       Cl   F

as long as it is electron deficient and inclined to accept an electron pair to form a covalent bond. Likewise, the Lewis base can have a full negative charge:

$$e.g. \quad Cl^{\ominus}, \quad CH_3O^{\ominus}, \quad CH_3^{\ominus} \quad \text{(negative ions)}$$

or a partial negative charge:

e.g. $-\overset{|}{C} - \underset{\delta-}{O}$ $\overset{H_{\delta+}}{\diagup}$ $\overset{\diagdown}{\underset{\delta+}{C}} = \underset{\delta-}{O}$ (polar bonds)

basic center

or be uncharged:

e.g. $CH_3\text{-}\ddot{N}H_2$, $CH_3 \overset{\ddot{O}\!\!:}{\diagup} \diagdown H$ (non-bonded electrons)

42

as long as it can donate a pair of electrons to a suitable acceptor to form a covalent bond. Here's a summary of these descriptions of Lewis acids and bases, which will be used frequently throughout this book:

LEWIS ACID	*LEWIS BASE*
electron pair acceptor	electron pair donor
electron-deficient species	electron-rich species
electrophile	nucleophile

The following figure (13-1) introduces a few important examples of how Lewis acids and bases are involved in the mechanisms of chemical reactions and physical interactions. The curved arrows in this figure trace the path of an electron pair as bonds form and break. All of these examples and many others, are described in greater detail throughout this book.

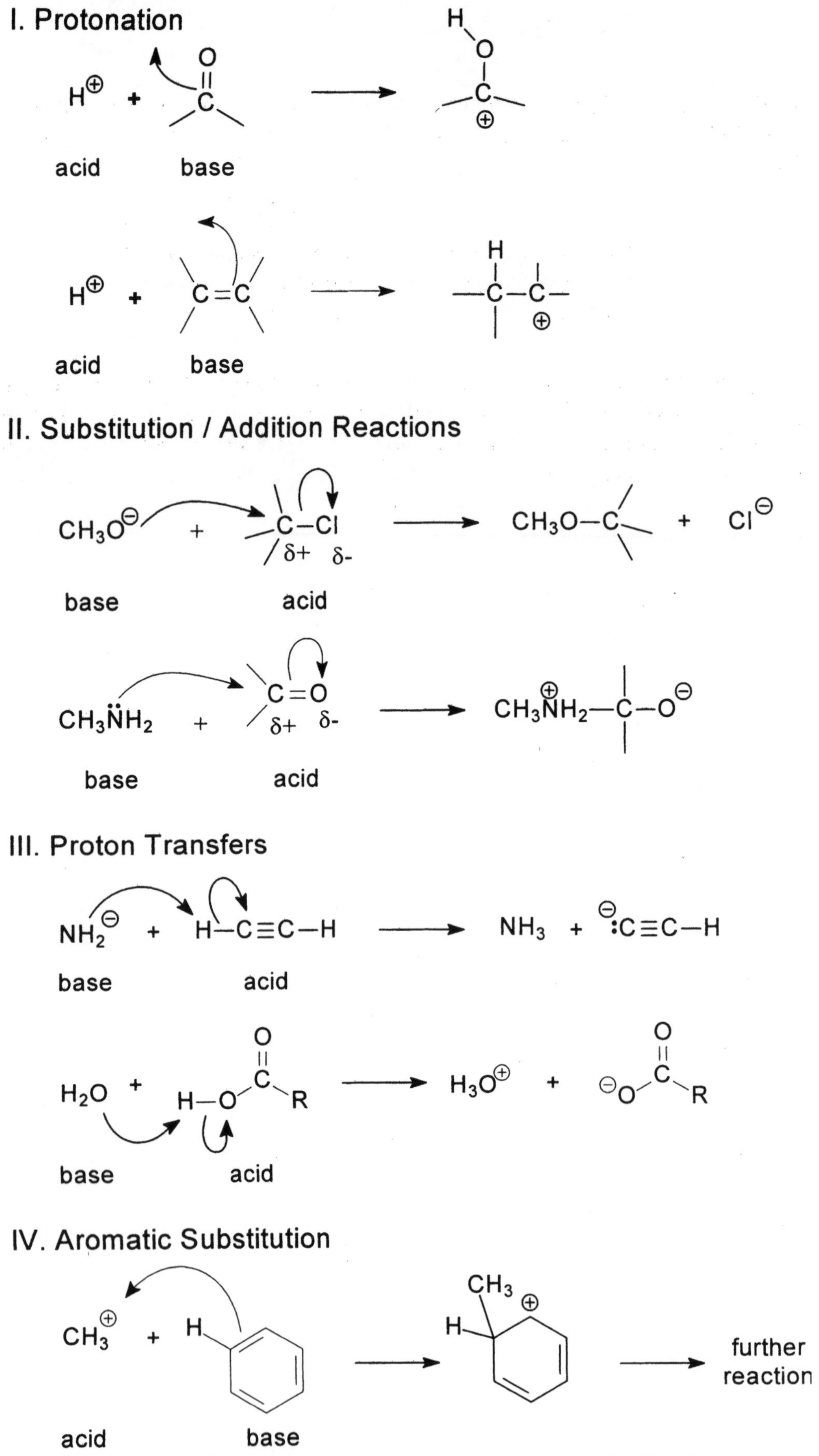
I. Protonation
acid
base
acid
base
II. Substitution / Addition Reactions
base
acid
base
acid
III. Proton Transfers
base
acid
base
acid
IV. Aromatic Substitution
further reaction
acid
base

Fig 13-1: Some examples of Lewis acids and bases in organic chemistry.

Chapter 14
Isomers and Stereochemistry

Isomers are different compounds having the same molecular formula. They possess the same number and kinds of atoms. Isomers can be generally classified into the broad categories of *constitutional isomers* and *stereoisomers*.

Constitutional Isomers

Constitutional isomers differ from each other by the order of connections between atoms. For example, two structures with the molecular formula C_2H_6O are ethyl alcohol and dimethyl ether. Each has the same number of carbon, hydrogen, and oxygen atoms,(2, 6, and 1, respectively), but they are attached in a different sequence, giving rise to molecules of very different physical properties and chemical reactivity.

Ethyl alcohol   Dimethyl ether

Another example of constitutional isomerism is the *propyl chlorides.* These compounds have the chlorine atom attached either to the central carbon atom or to a terminal carbon atom. These isomeric forms are different compounds with slightly different physical properties like boiling point and refractive index, and exhibit different rates of chemical reactivity as well:

$$Cl-CH_2-CH_2-CH_3 \qquad CH_3-\underset{\underset{Cl}{|}}{CH}-CH_3$$

1-Chloropropane   2-Chloropropane

Another example of constitutional isomers are the various *dichlorobenzenes:*

ortho-
dichlorobenzene

meta-
dichlorobenzene

para-
dichlorobenzene

45

Here, the chlorine atoms are attached to adjacent, alternate, or opposed ring carbon atoms, producing three different compounds - called ortho, meta, and para isomers. In similar fashion, the site of attachment of the methyl groups on the same carbon atom or on neighboring carbons gives rise to the 1,1 or 1, 2 constitutional isomers of *dimethylcyclopropane.*

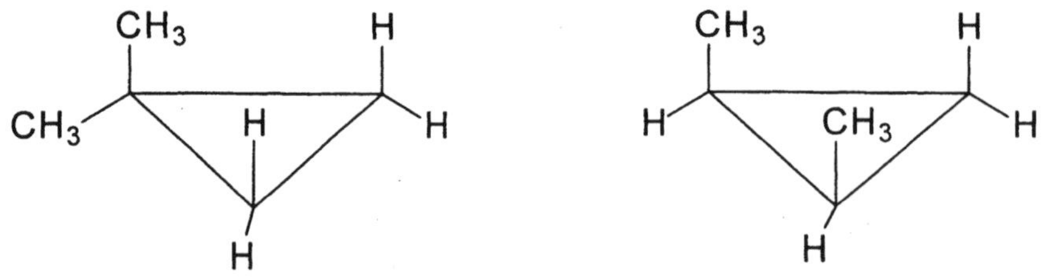

1,1-dimethylcyclopropane   *cis*-1,2-dimethylcyclopropane

Isomers are ubiquitous throughout organic chemistry, and are a consequence of the huge diversity of structures made possible by the covalent bonding ability of carbon atoms. Indeed, certain types of molecules composed of only 20 carbon atoms can theoretically be connected differently to form over a third of a million different isomers!

Stereoisomers:

Another type of isomerism is *stereoisomerism.* These isomers differ from each other based on the *shape* or *arrangement in space* of the molecule, while the order of connectivity remains the same. For example, the 1,2 dimethylcyclopropanes can have the methyl groups on the same or on opposite sides of the plane of the three-membered ring. These are called *cis* and *trans* isomers respectively:

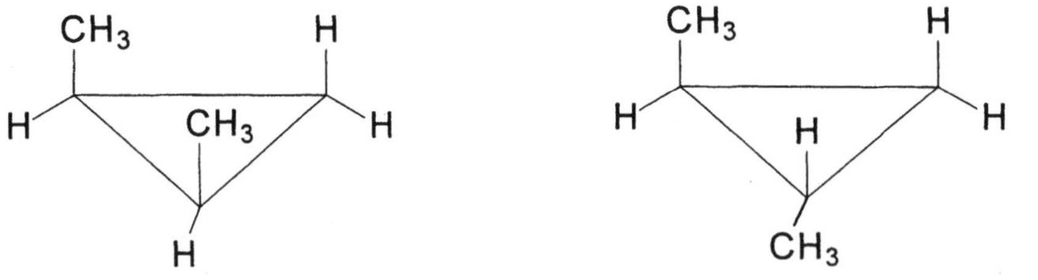

cis-1,2-dimethylcyclopropane   *trans*-1,2-dimethylcyclopropane

There is no way these isomers can interconvert under normal circumstances, since that would involve either rotation through the center of the rigid three-membered ring, which is not possible, or would require breaking, rotating, and then reforming a carbon-carbon bond of the ring atoms, which would also require too much energy. In *open-chain, non-cyclic* structures having only carbon-carbon *single* bonds, cis-trans isomerism generally does not occur, since only small barriers to free rotation exist in such compounds.

46

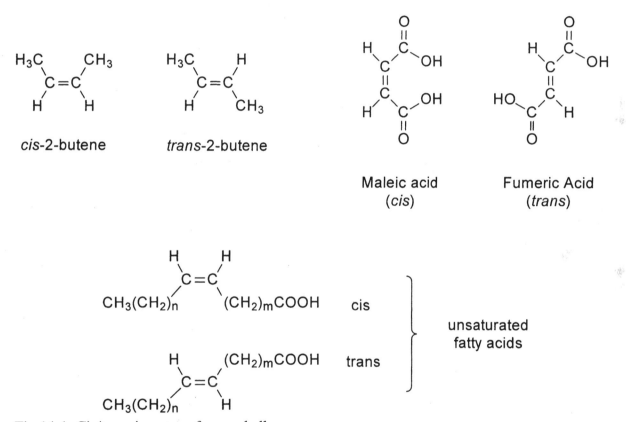

There is little barrier to free rotation in most open chain compounds.

Suitably substituted *olefins* (alkenes), which have carbon-carbon *double* bonds, also exist as cis-trans isomers. In this case, inter-conversion would necessitate rotation about the double bond. This rotation can occur only if the pi bond is broken, which is not possible in normal circumstances due to the very high energy required to break this bond.

cis-2-butene   trans-2-butene   Maleic acid (*cis*)   Fumeric Acid (*trans*)

cis

trans

unsaturated fatty acids

Fig 14-1: Cis/trans isomers of several alkenes.

Optical Isomers

Another very interesting and important type of stereoisomerism is called *optical isomerism*. Optical isomerism derives from the fact that a tetrahedral (saturated) carbon atom, when attached to *four different* atoms or groups of atoms, has the same symmetry relationship to its mirror image as our left and right hands have to each other. They are both *non-superimposible* mirror images. Compounds having this structural feature are said to be *chiral*, which means *handed*. The carbon atom directly attached to the four different groups is called an *asymmetric center* or an *asymmetric carbon*, and the isomeric mirror images they form are called *enantiomers, racemates,* or *optical isomers*. The term optical isomer derives from the fact that a solution of one of

47

the mirror image enantiomers of such a compound can rotate the plane of polarized light to the right while the other mirror image form rotates polarized light to the left. An instrument called a *polarimeter* is used to carry out and measure this rotation. Enantiomers share the same physical and chemical properties; they differ from each other only in their ability to rotate the plane of polarized light in one direction or the other.

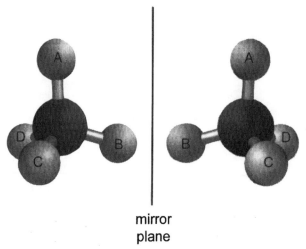

mirror
plane

Fig 14-2: Enantiomers. To be non-
superimposible mirror images, A, B, C, and D,
must all be different atoms or atomic groupings.

Most important biological molecules have asymmetric centers and exist in nature in only one of the two possible handed (enantiomeric) forms. These include all of the amino acids (except glycine) and the proteins resulting from their linkage, and all of the simple sugars and the carbohydrates formed from them.

$$
\begin{array}{ccc}
\mathrm{COOH} & \mathrm{COOH} & \\
| & | & \\
\mathrm{R-\overset{*}{C}-H} & \mathrm{H-\overset{*}{C}-R} & \text{Amino acids} \\
| & | & \\
\mathrm{NH_2} & \mathrm{NH_2} &
\end{array}
$$

$$
\begin{array}{ccc}
\mathrm{CHO} & \mathrm{CHO} & \\
| & | & \\
\mathrm{H-\overset{*}{C}-OH} & \mathrm{HO-\overset{*}{C}-H} & \text{Glyceraldehyde} \\
| & | & \\
\mathrm{CH_2OH} & \mathrm{CH_2OH} &
\end{array}
$$

Fig 14-3: Enantiomeric biological molecules. Only one chiral (handed) form generally exists in nature. (The asterix indicates an asymmetric carbon atom.)

48

An enantiomer is identified as (d) or (+), and is called *dextrorotatory* if it rotates the plane of polarized light to the right (clockwise). The mirror image enantiomer, which rotates the plane of polarized light to the left, (counterclockwise) is called the (l) or (−) or *levorotatory* form. Somewhat confusingly, the *capital letters* D and L are designations formerly used to identify the actual 3-dimensional spatial orientation of these isomers, which does not relate in a simple way to the direction of rotation of polarized light. In recent times, these designations have been replaced with the capital letters R and S, which stand for *Rectus* and *Sinister*. These capital letters are used to describe the spatial orientation of each asymmetric carbon in a chiral molecule, *not* the direction of rotation of plane polarized light. (If there are several asymmetric carbons in a molecule, each must be separately identified as R or S.)

Compounds that have *more than one* asymmetric center and that are not mirror images of each other *do* have different physical and chemical properties, such as melting points, boiling points, solubility, and rates of chemical reactions. These stereoisomers are called *diastereomers,* and are exemplified by the drugs

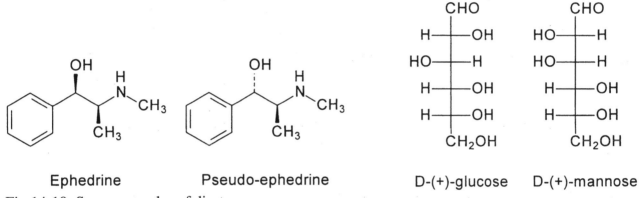

Ephedrine   Pseudo-ephedrine   D-(+)-glucose   D-(+)-mannose

Fig 14-10: Some examples of diastereomers

ephedrine and pseudoephedrine, and the sugars glucose and mannose. *Diastereomers, in general, are defined as stereoisomers that are not mirror images of each other.*

Compounds with multiple asymmetric carbons will have, at most, a total number of isomeric forms given by the expression 2^n, where n is the number of asymmetric carbon atoms in the molecule. Thus, the

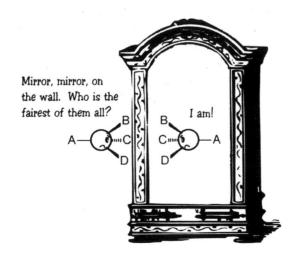

Mirror, mirror, on the wall. Who is the fairest of them all?

I am!

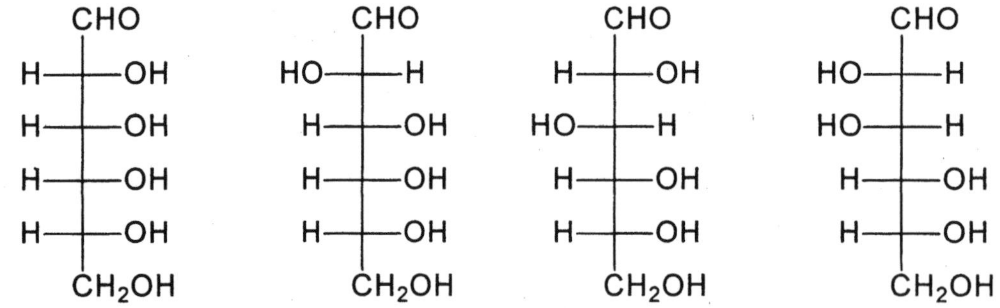

Fig 14-5: The diastereomers of the hexose sugars.

D-(+)-allose, D-(+)-altrose, D-(+)-glucose, D-(+)-mannose, 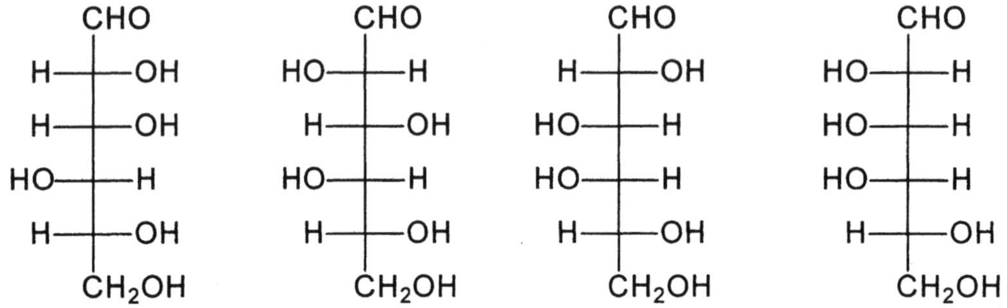 D-(-)-gulose, D-(+)-idose, D-(+)-galactose, D-(+)-talose

The following overview, discussion, and examples will help to clarify our understanding of optical isomerism and chirality:

1. An excess of one or the other enantiomeric forms of a compound is required to rotate the plane of polarized light. A *racemic mixture,* which by definition is composed of equal quantities of each form (a 50/50 mixture) will *not* rotate the plane of polarized light, since the clockwise rotation caused by one form is exactly cancelled by the equal counterclockwise rotation caused by the other. Certain compounds with asymmetric centers, called *meso compounds,* have an *internal plane of symmetry,* in which half of the molecule is the mirror image of the other half. Meso compounds do not exhibit net rotation of polarized light due to the equal but opposite rotations induced by the two halves of the molecule. A *pure enantiomer* (optical isomer) is composed entirely of only one mirror image form. It is an interesting (and poorly understood) fact that virtually all biological molecules containing asymmetric carbons occur naturally in only one of the two possible mirror image forms.

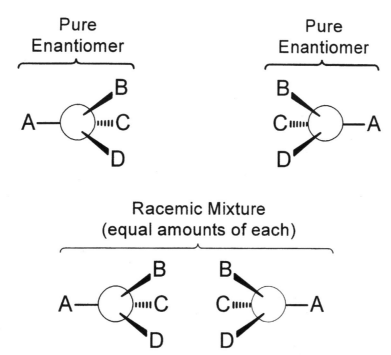

Fig 14-6: Enantiomers and a racemic mixture.

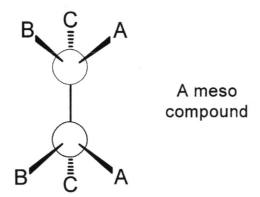

A meso compound

Fig 14-7: A meso compound. The top half of the molecule is the mirror image of the bottom half.

2. Two or more asymmetric carbons in a molecule can produce diastereomers having physical and chemical properties that differ in every way. This is easily understood by a simple analogy: shaking hands. The meeting of two *right* hands results in a good "fit" and a successful handshake; attempting to shake right and left hands results in a poor "fit" and a failed handshake. Similarly, in

molecules with more than one asymmetric center, each having the same symmetry properties as our hands: the spatial interaction between the centers differs depending on the handedness of the two sites. In fact, the asymmetric centers need not be on adjacent carbons within a molecule or in the same molecule to produce these differences, as long as they are in close proximity. Thus, *enzymes,* having an asymmetric active site, may catalyze the reaction of one enantiomer, but not the other. *Antibodies* often bind to one enantiomeric *antigen* but not the other, since proper fit is needed for this interaction to occur as well, and diastereomeric salts, where the anion and cation are each pure enantiomers, often have very different solubilities and other physical properties, as long as they are not mirror-images of each other. This is the basis of the classic method of separating (resolving) enantiomers by fractional crystallization of diasteriomeric salts.

3. Chemical reactions of pure enantiomers can proceed with retention, inversion, or loss of optical activity. This allows for the design of experiments that can shed light on the mechanism of chemical reactions by studying the optical rotation of the products. For example, *inversion of configuration* will occur if a substitution reaction occurs by concerted "backside attack" by the substituting group:

Fig 14-8: Inversion of configuration in a substitution reaction.

On the other hand, *loss* of optical activity would occur if a *planar* intermediate is formed, since this would bond to the substituting group, N⁻, with *equal likelihood from either side.* Reaction by this mechanism would produce a racemic product mixture composed of equal quantities of the two mirror image (enantiomeric) forms, which would *not* rotate the plane of polarized light in a polarimeter.

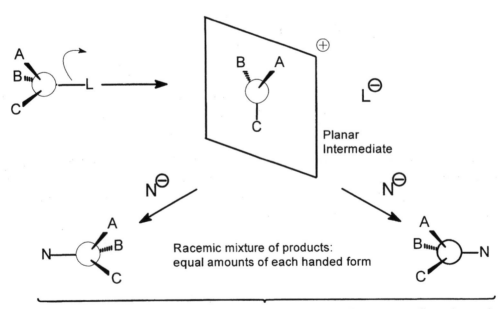

Fig 14-9: Loss of optical activity in a substitution reaction proceeding through a planar intermediate.

Chapter 15
Introduction to Organic Reactions

Except in very unusual circumstances, electrons in stable organic compounds always occur in pairs. A pair of electrons is represented by two dots or more commonly by a line segment connecting two bonded atoms. *One line is always the equivalent of a two-electron bond.* Non-bonded electron pairs are usually indicated by dots:

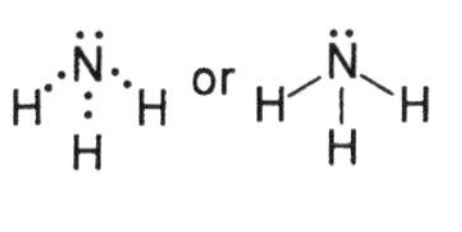

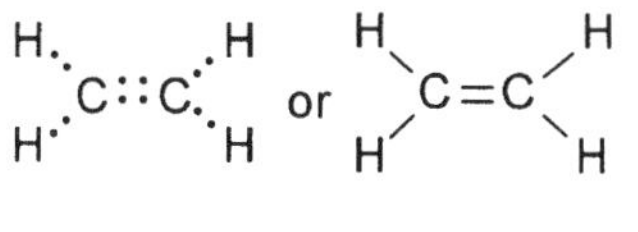

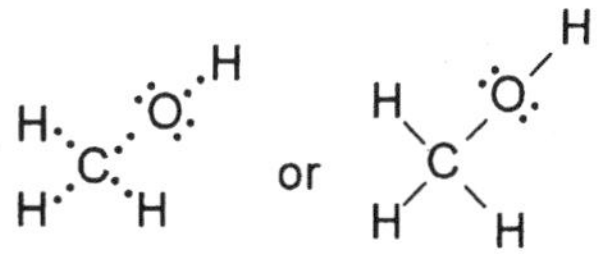

The breaking of these two-electron bonds in the course of a chemical reaction is generally indicated by curved arrows. If the bond breaks so that one atom gets *both* electrons and the other gets *neither,* the bond cleavage is called *heterolytic.* This results in the formation of oppositely charged ions from a single neutral molecule, as in the example below. If a carbon atom carries the positive charge, the ion is called a *carbocation* or *carbonium* ion. This is an unstable, reactive intermediate, which always reacts further to re-acquire an electron octet.

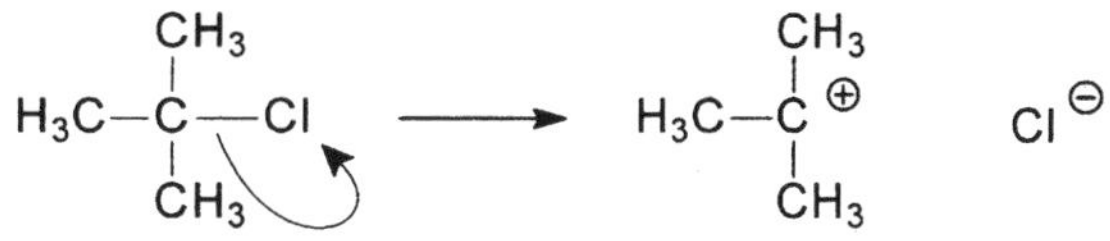

The two-electron pi bond of an alkene can also break to form a sigma bond to a proton, H^+. The proton can bond to either the left or right carbon of the alkene, as indicated by the direction of the curved arrow which shows the movement of the pi electron pair. Both electrons are contributed by the alkene which is thus the Lewis base; the proton is the Lewis acid. As in the example above, the resulting unstable carbocation then reacts further to re-acquire an electron octet.

carbocation

or

carbocation

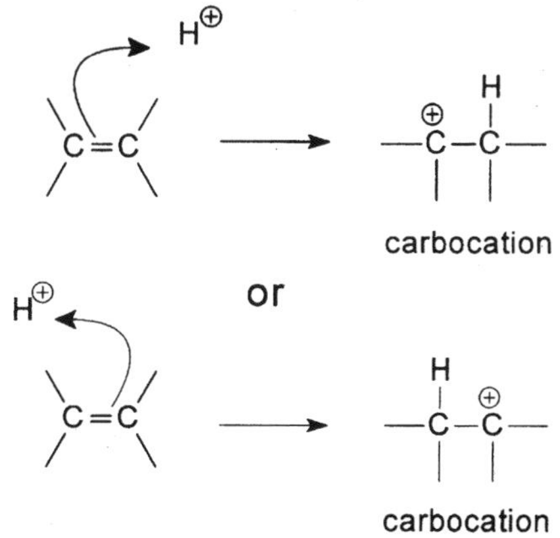

Bond cleavage can also occur to leave the electron pair and the negative charge on the carbon atom, as in the acid-base reaction between the amide ion, NH_2^-, of sodium amide, and an alkyne. A carbon atom carrying a negative charge is called a *carbanion.* The product in this case is the uncharged ammonia molecule and the carbanion derived from the alkyne. The direction of the curved arrows indicated the flow of electron pairs. (The spectator Na^+ ion is not shown in this example, as is common practice.)

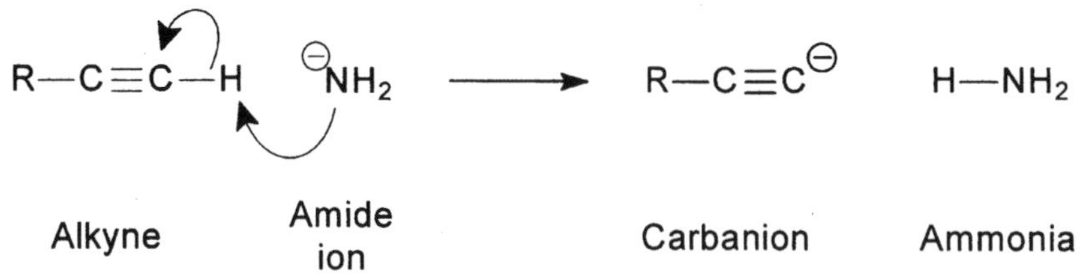

| Alkyne | Amide ion | | Carbanion | Ammonia |

Arrows are also used to indicate a *homolytic* cleavage in which each fragment retains one of the two formerly bonded electrons. This mode of cleavage results in uncharged reactive fragments called *free radicals.* This is shown below for several steps involved in the polymerization of ethylene.

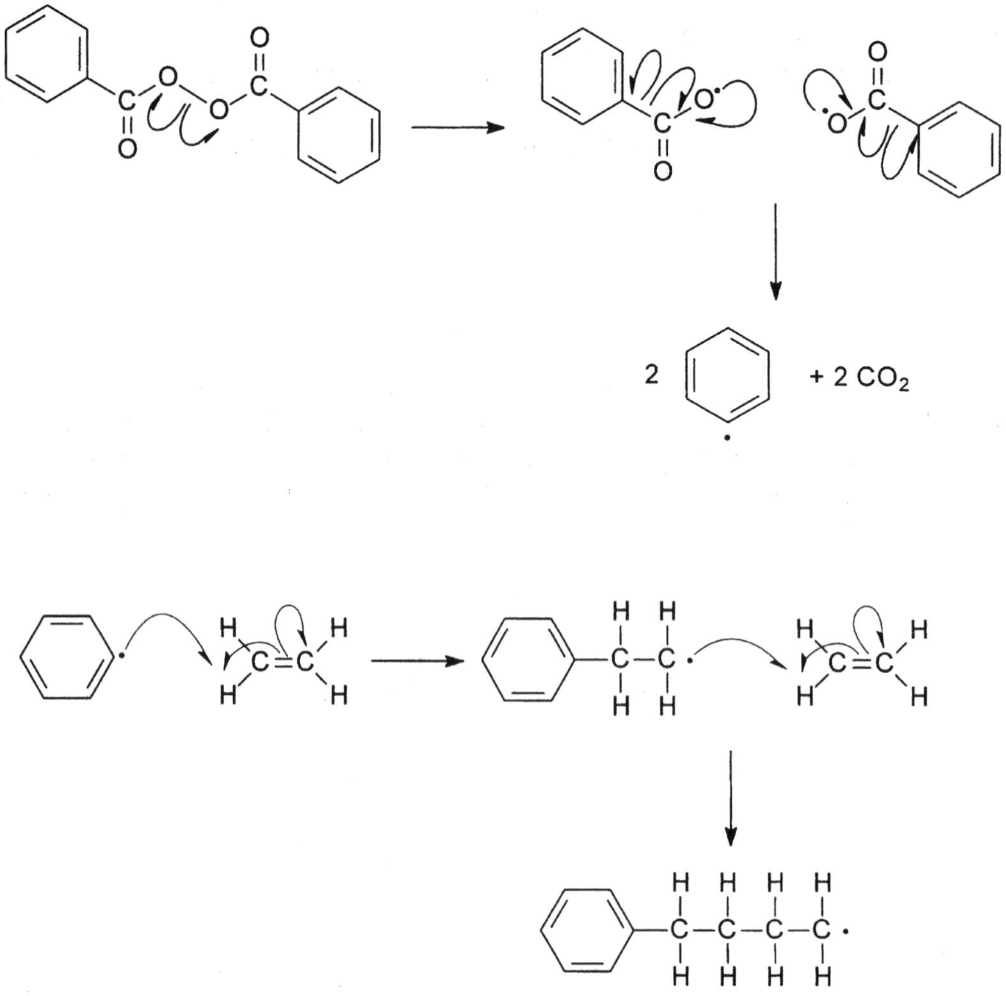

$$2 \bigcirc \cdot \quad + \ 2\ CO_2$$

The arrows shown in the preceding examples are employed to help our eyes follow the process of bonds breaking and forming in a chemical reaction. Arrows are also a teaching and bookkeeping aid to make sure we keep track of all the valence electrons as a reaction proceeds. The principles of conservation of charge and mass demand that all electrons always be accounted for in any chemical reaction.

Complicated reaction mechanisms must still follow these same simple rules. The curved arrows in the example below help us keep track of the flow of electrons as the reaction unfolds.

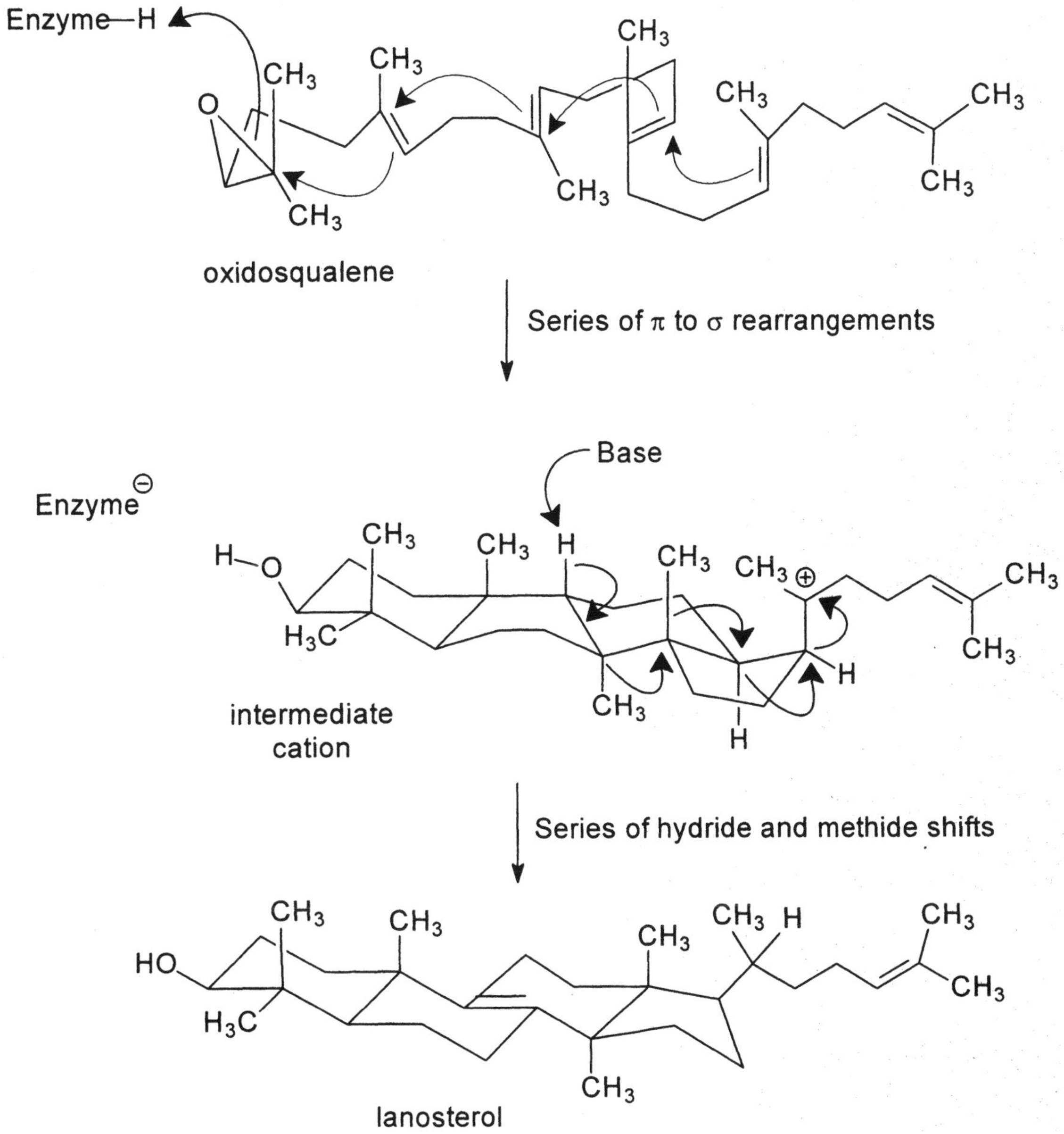

Enzyme—H
CH3
O
oxidosqualene
Series of π to σ rearrangements
Base
Enzyme⊖
H—O
H3C
H
intermediate
cation
Series of hydride and methide shifts
HO
lanosterol

For a slightly different example, consider a step in the *Grignard reaction,* shown below. Here, two electrons are transferred from a magnesium atom to an alkyl halide. The two-electron carbon-chlorine sigma bond of the alkyl halide then breaks. Each fragment retains two of the four available electrons producing a carbanion, R^-, and a chloride ion, Cl^-. (The reaction product $R^- Mg^{2+} Cl^-$ must have a net charge of zero since the net charge on *any* completely described covalent or ionic compound is always zero.)

$$\text{R–Cl} + \text{Mg} \xrightarrow{2e^-} \left[(\text{R–Cl})^{-2} \text{Mg}^{+2} \right] \longrightarrow \text{R}^{\ominus}\text{Mg}^{+2}\text{Cl}^{\ominus}$$

A single electron can also be transferred from each of two lithium atoms in a similar reaction:

$$\text{R–Cl} + 2\text{Li} \xrightarrow{2e^-} \left[(\text{R–Cl})^{-2} \, 2\,\overset{\oplus}{\text{Li}} \right] \longrightarrow \overset{\oplus}{\text{Li}}\text{R}^{\ominus} + \overset{\oplus}{\text{Li}}\text{Cl}^{\ominus}$$

Here are several more examples of the use of arrows to help describe and clarify reaction mechanisms.

Dashed or dotted lines are often used to indicate *partial* bonds which are in the *process* of being broken or formed, but have not completely done so. This condition is referred to as a *transition state,* sometimes abbreviated TS. For example:

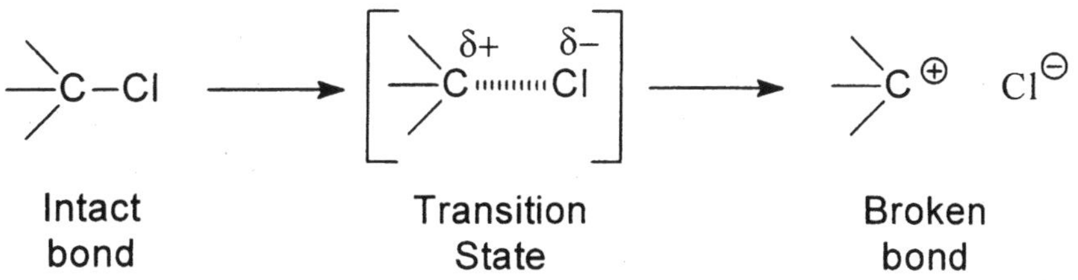

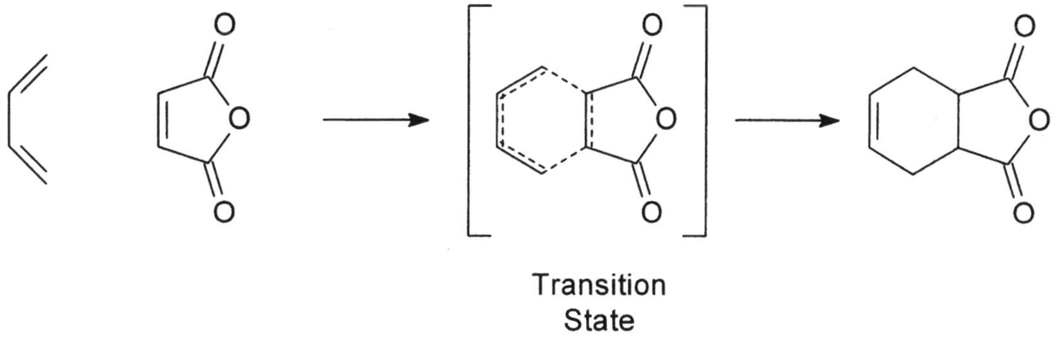

Intact
bond

Transition
State

Broken
bond

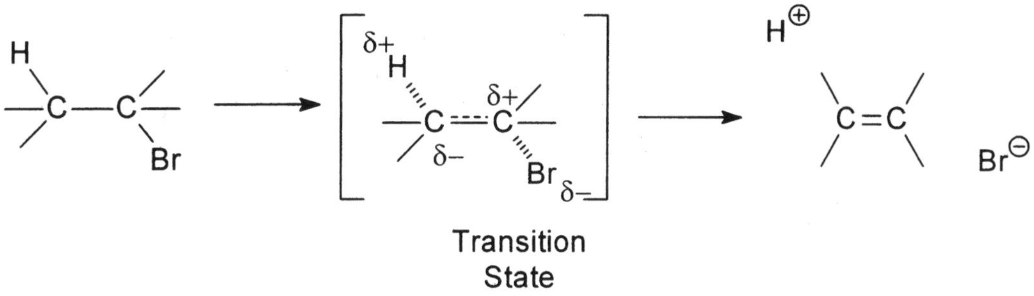

Transition
State

Transition
State

Chapter 16
Organic Reaction Mechanisms: A General Overview

Reaction mechanisms describe the detailed sequence by which chemical bonds are broken and formed in a reaction. In organic chemistry only a few simple mechanistic steps commonly occur in describing a great many different reactions. This section will briefly present an overview of these mechanisms; later sections will describe mechanisms in much more detail as specific reactions are discussed.

One-Step Mechanisms

In some cases bonds are broken and formed *simultaneously*. Such reactions are said to proceed through a *concerted, simultaneous,* or *one-step* mechanism. A few common examples of this type of reaction mechanism are shown below. These include the *Diels-Alder* reaction and (bimolecular) substitution and elimination at a saturated carbon atom:

Diels-Alder Reaction

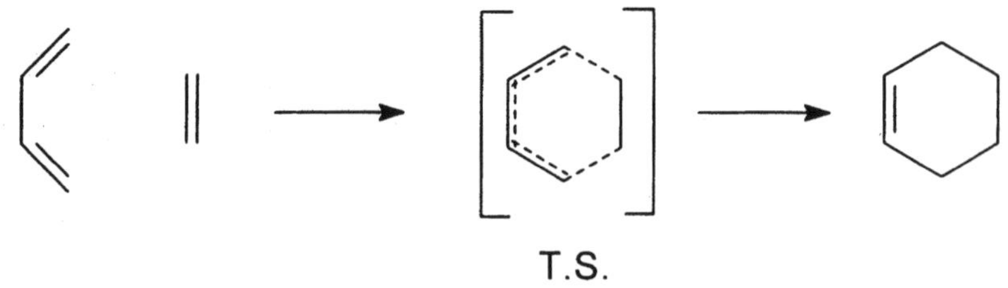

T.S.

S$_N$2 Substitution

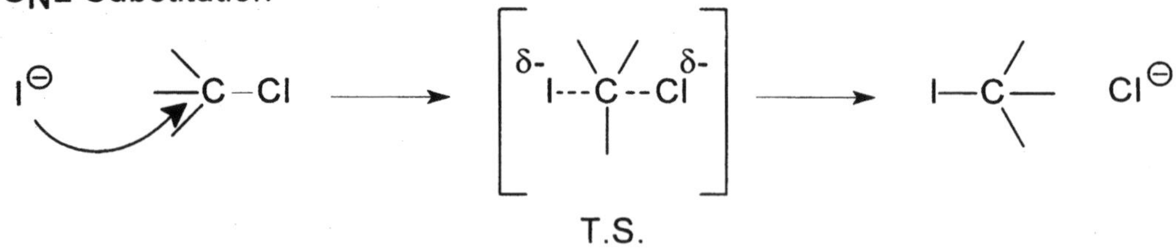

T.S.

E2 Elimination

T.S.

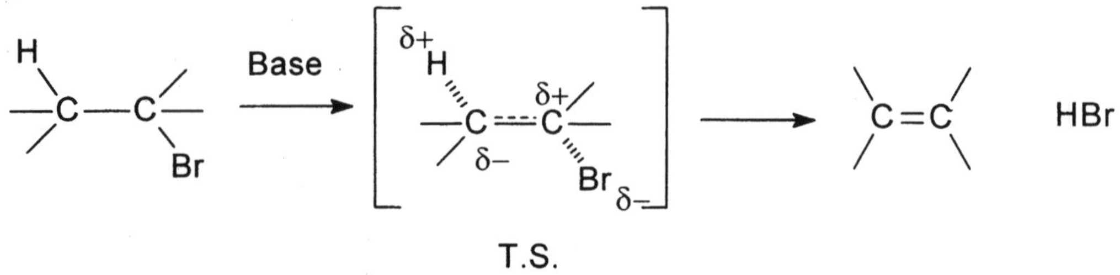

The dotted lines in the Transition State (T.S.) in brackets above, indicate *partial* chemical bonds. Partial bonds occur as bonds are breaking in the starting compounds and as they are simultaneously forming in the reaction products. The Transition State represents the maximum energy barrier which the reactants must surmount for the reaction to go on to completion and form stable products.

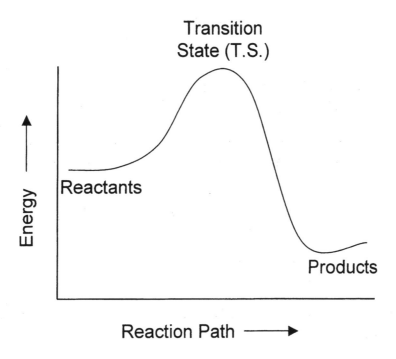

Multi-Step Mechanisms and Reactive Intermediates

Chemical bonds can break and new bonds can form in a stepwise fashion as well. Such reactions are said to proceed by a *stepwise, multi-step,* or *sequential* mechanism. Here, bonds are initially broken to form fragments called *reactive intermediates,* which, though unstable, do have a finite lifetime and independent existence. These reactive intermediates then react further to form stable products. Only a few types of reactive intermediates exist, the most important being *free radicals,* which are uncharged; *carbanions,* which have a negative charge; and *carbocations* (also called *carbonium ions*), which have a positive charge.

Free Radicals

In organic compounds atoms are linked together by covalent bonds. Each bond is composed of two electrons. If one of these bonds cleaves "down the middle," so each fragment retains one electron, the cleavage is said to be *homolytic* and the resultant fragments, called *free radicals,* are electrically neutral. The unpaired electron carried by the cleaved fragments is generally indicated by a "dot"($\cdot$). Free radicals are unstable and highly reactive since they generally have only 7 valence shell electrons. They rapidly engage in further reactions which eventually allow them to re-establish the stable octet.

59

A common example of a free-radical reaction is the chlorination of saturated hydrocarbons, like methane. In this reaction, chlorine gas is irradiated with UV(ultra-violet) light to cleave the molecule into two atoms (free radicals):

$$Cl_2 \xrightarrow{\text{UV light}} 2\ Cl\bullet$$

Upon collision with a hydrocarbon like methane, these chlorine atoms will then react to abstract (pull off) a H· atom forming HCl and a methyl radical (·CH$_3$). The methyl radical then reacts with a molecule of Cl$_2$ to produce the product, methyl chloride, and another Cl· atom. These steps (a) and (b) repeat through many cycles to yield many molecules of methyl chloride. This is called a *chain reaction* since a new reactive chlorine atom, Cl·, is produced each time the 2-step sequence (a and b) is completed.

(a) $Cl\bullet\ +\ CH_4 \longrightarrow HCl\ +\ \bullet CH_3$

(b) $\bullet CH_3\ +\ Cl_2 \longrightarrow CH_3Cl\ +\ Cl\bullet$

Eventually the chain is terminated by the combination of any two radicals to form a stable product having 8 valence electrons around each atom:

$$2\ Cl\bullet \longrightarrow Cl_2$$

$$\bullet CH_3 + Cl\bullet \longrightarrow CH_3Cl$$

$$2\ \bullet CH_3 \longrightarrow CH_3CH_3$$

Charged Intermediates

While some important reactions proceed through free-radical intermediates, by far the majority of reactions involve ionic (charged) intermediates. These carry either a positive charge (carbocations) or a negative charge (carbanions).

Carbocations (Carbonium ions)

If a covalent bond breaks so that one fragment retains both electrons and the other retains neither, this is referred to as a *heterolytic cleavage:*

$$A\text{---}B \xrightarrow[\text{cleavage}]{\text{heterolytic}} A^{\oplus} +\ B^{\ominus}$$

If the cleavage occurs at a carbon atom with the other fragment retaining the electron pair, the positive carbon species is called a *carbonium ion* or *carbocation*. For example:

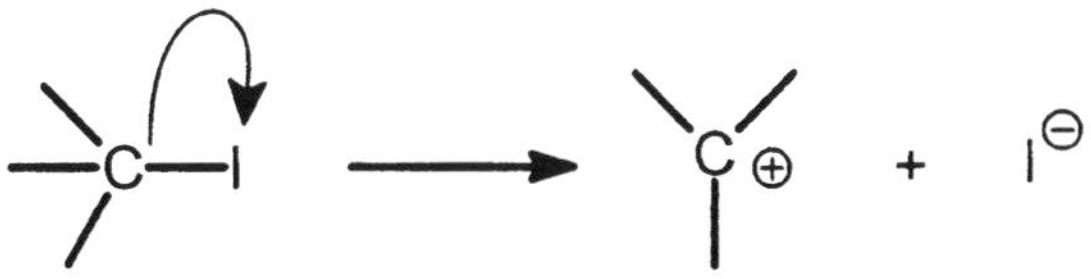

carbocation

Carbocations also result if a positive species, e.g., a proton, forms a bond with a neutral organic molecule. For example:

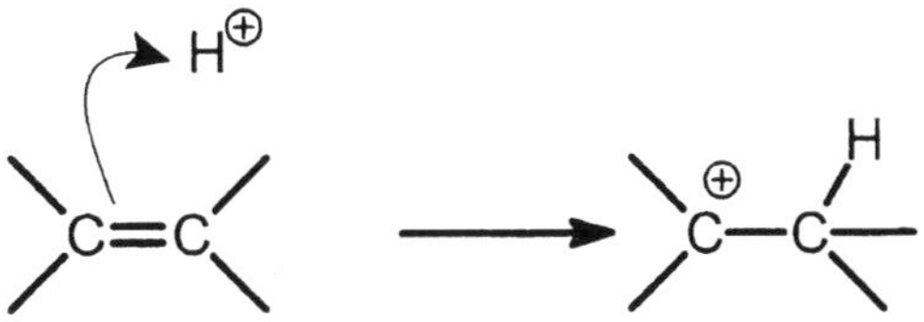

The carbocation (carbonium ion) is a highly reactive intermediate, since the carbon atom has only 6 valence electrons. It rapidly reacts further to achieve the desired octet. Carbocations are very strong Lewis acids (electron pair acceptors) and for this reason react avidly and rapidly with Lewis bases (electron pair donors) to form new covalent bonds. They can also reacquire an octet by proton loss. Carbocations are very important intermediates. They are involved in many reactions including addition, elimination, and substitution. For example:

1. Addition

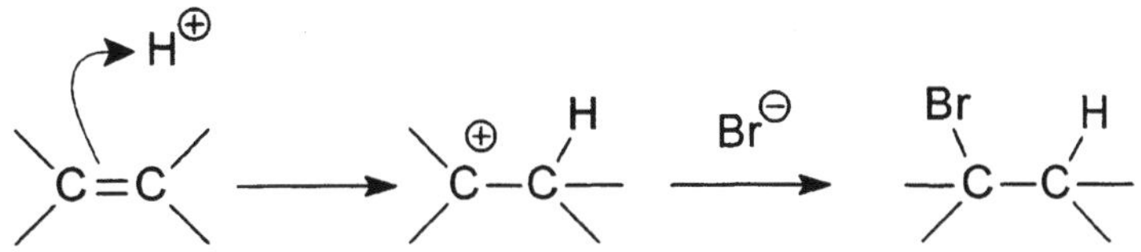

2. Elimination

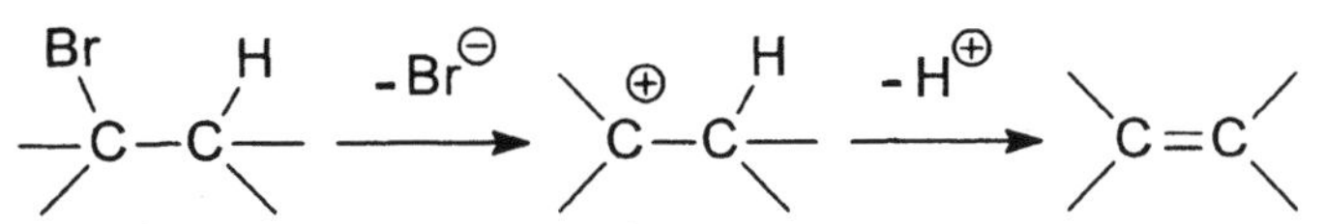

3. Substitution

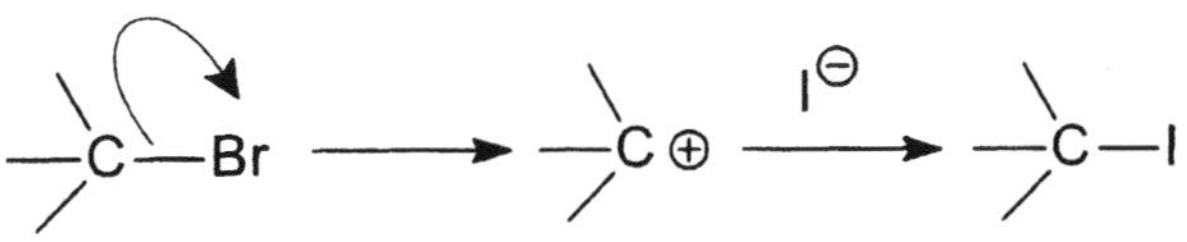

Fig 16-1: Some important carbocation reactions.

61

Carbanions

Another important class of intermediates involves carbon atoms bearing a *negative* charge. This reactive species is called a *carbanion.* Carbanions can form if a hydrocarbon, RH, behaves as an acid and liberates a proton by reacting with a suitably strong base:

$$R-H + Base \longrightarrow R^{\ominus} + BaseH^{\oplus}$$
$$\text{carbanion}$$

Most hydrocarbons, however, are too weakly acidic to undergo this acid-base reaction. More commonly, carbanions are produced by an electron transfer reaction between an active metal, most often magnesium, and an alkyl halide, RX, to form a mixed salt containing a carbanion.

$$R-X + Mg \longrightarrow \left(R^{\ominus}Mg^{2\oplus}X^{\ominus} \right)$$

Carbanions react as very strong Lewis bases and as strong *nucleophiles,* attacking and forming bonds to electron deficient centers (Lewis acids). For example:

$$R^{\ominus} \quad \underset{\delta+ \ \ \delta-}{C=O} \longrightarrow R-C-O^{\ominus}$$

$$R^{\ominus} \quad \underset{\delta+ \ \ \delta-}{C-Cl} \longrightarrow R-C + Cl^{\ominus}$$

As strong bases, they also react avidly with a wide variety of proton sources:

$$R^{\ominus} \quad H-A \longrightarrow RH + A^{\ominus}$$

Note that carbanion reactivity derives from the fact that they are strong bases and nucleophiles, while the reactivity of radicals and carbocations is due to the fact that they both lack an octet of valence shell electrons.

Other Intermediates

Other reactive intermediates are known in organic chemistry. These can arise by electron loss or electron gain of a molecule, M, to produce *radical-cations* or *radical-anions.* These electron transfer reactions are important in *oxidation-reduction* reactions.

$$M - e^{\ominus} \longrightarrow M^{+}_{\bullet} \quad \text{Radical Cation}$$

$$M + e^{\ominus} \longrightarrow M^{-}_{\bullet} \quad \text{Radical Anion}$$

Finally a neutral species called a *carbene* is involved in the mechanism of a few reactions.

$$\left[R{-}\overset{..}{C}{-}R \right] \quad \text{Carbene}$$

SUMMARY

Reaction mechanisms describe the details of the bond-breaking and bond-forming steps which convert reactant molecules into product molecules. A small number of simple mechanistic steps are common to a great many different organic reactions. These events can happen simultaneously or stepwise. Stepwise mechanisms proceed through unstable, highly reactive intermediates which can either be electrically neutral (free radicals) or possess a positive or negative charge (called carbocations and carbanions, respectively). These reactive intermediates then react further to eventually yield stable reaction products.

Free Radical

Chapter 17
Steric Hindrance

The principle of *steric hindrance* is one of the simplest and most readily understood concepts in all of science. It says that chemical events tend to occur with special difficulty, or not at all, if two atoms or atomic groups are required to *occupy the same space simultaneously.* This results in the intrusion of one electron cloud into the space occupied by another; the resulting negative charge repulsion raises energy and decreases stability. Steric hindrance has consequences for chemical reactivity: slowing down reactions involving highly hindered transition states, or causing reactions to speed up when steric strain is relieved as the reaction proceeds. The relative stability or instability of various constitutional isomers, stereoisomers, and rotational conformations can also be explained on this basis. The following examples show how the concept of steric hindrance is used to describe and explain a variety of chemical and physical properties of organic compounds.

1. Rotational Conformation

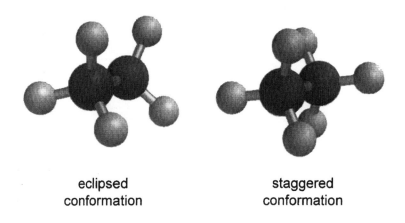

eclipsed
conformation

staggered
conformation

The staggered conformation is preferred to increase
separation of attached groups

Fig 17-1: Steric hindrance in organic chemistry.

2. Reaction Rates

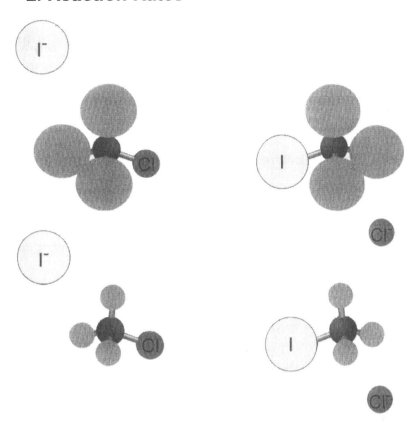

The large "bulky" groups block the approach of the iodide ion, thus retarding substitution for chloride by this mechanism.

Fig 17-1 (continued)

3. Molecular Shape

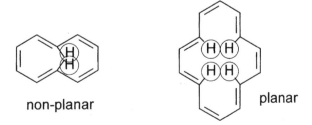

non-planar   planar

The smaller cyclic molecule on the left cannot achieve planarity due to steric hindrance of the interior H atoms. The structure on the right is large enough to accommodate the interior hydrogens, and is planar.

Fig 17-1 (continued)

4. Thermodynamics

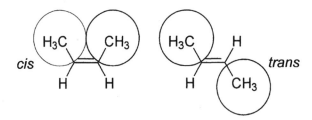

Combustion of the *cis* isomer releases slightly more heat (energy) than does the *trans* isomer because of the lower stability of the former due to crowding of the methyl groups.

Fig 17-1 (continued)

5. Product Isomer Distribution

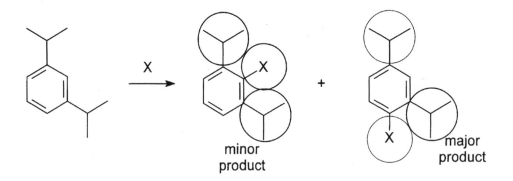

Little of the minor product is formed due to excessive steric hindrance in the transition state leading to that product.

Fig 17-1 (continued)

Backseat driving: major cause of steric hindrance.

66

Part 2
THE FUNCTIONAL GROUPS: PROPERTIES AND REACTIVITY
Chapter 18
The Alkane Hydrocarbons

The simplest class of organic compounds are the *alkane hydrocarbons.* These compounds contain only carbon and hydrogen, with all carbon atoms linked only by single bonds. This class of compounds is also referred to as *saturated hydrocarbons* because the carbon atoms are bound to the maximum possible number of other atoms (four); that is, the carbons are *saturated* with bonded atoms. Finally, they are also called *paraffins,* a name which derives from the low level of chemical reactivity of these compounds. The names and various representations of a few of the simplest members of this class of compounds are shown in the table below:

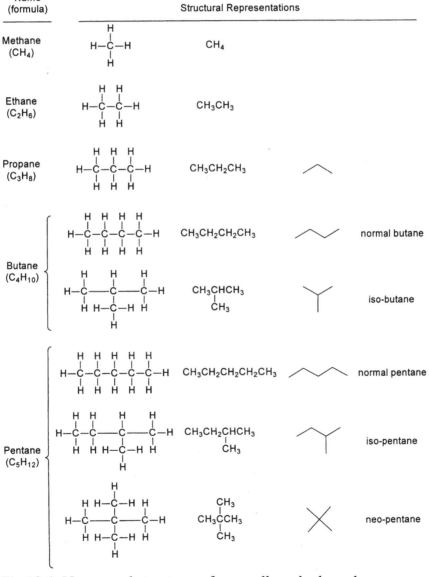

Fig 18-1: Names and structures of some alkane hydrocarbons.

Alkanes with 1, 2, 3, and 4 carbon atoms have only historical (non-systematic) names: methane, ethane, propane, and butane, respectively. Compounds having more than four carbon atoms have names that begin with the Greek prefix identifying the number of carbon atoms; *pent-, hex-, hept-,* and *oct-,* and so on. Many compounds have both historical (common) names and systematic names. The systematic names follow a set of rules developed at a series of international meetings convened for that purpose. Both sets of names should be learned.

Several representations of each compound are given in the figure on the previous page. The simplest is the *molecular formula,* which describes only the *number* of carbon and hydrogen atoms in the molecule, for example, C_3H_8. Three types of structural formulas are shown as well, which describe the *connectivity* of each atom to its neighbors. One version explicitly shows the linkage of all bonded atoms. For example, for iso-pentane:

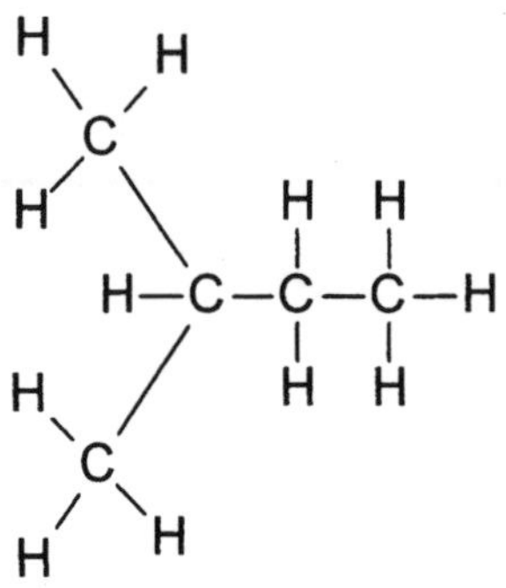

Another, called a *condensed* formula, represents groups or clusters of bonded atoms;

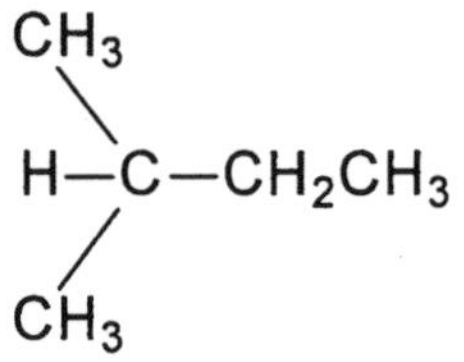

And the simplest shorthand version, called a *bond-line* representation, indicates the structure as a line figure, like this:

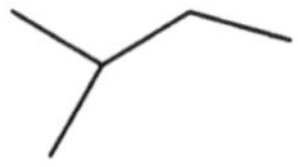

In the bond-line figure, it is understood that a carbon atom is present at all points where the lines begin, end, or change direction; and it is further understood that each carbon is attached to the number of hydrogen atoms required to complete the octet around that carbon atom.

Each of these representations is an unambiguous description of the same molecule. For alkanes exceeding 3 carbon atoms, the carbons can be linked (connected) together in various ways. These structures, which all have the same molecular formula but a different order of connection, are called *constitutional isomers* of each other. Thus, for pentane, C_5H_{12}, three different isomers exist, as seen in the figure above. The number of possible linkages increases, of course, with the number of carbon atoms in the formula. Seventy-five isomeric possibilities exist for a 10-carbon alkane ($C_{10}H_{22}$) while, incredibly, over a third of a million isomeric possibilities exist for the 20 carbon formula, ($C_{20}H_{42}$). You may have already noted that all the open chain (non-cyclic) alkanes conform to the generalized molecular formula of C_nH_{2n+2}.

Cyclic, or closed chain alkanes also exist. Here are three representations of the closed ring (cyclic) structure, cyclohexane, C_6H_{12}.

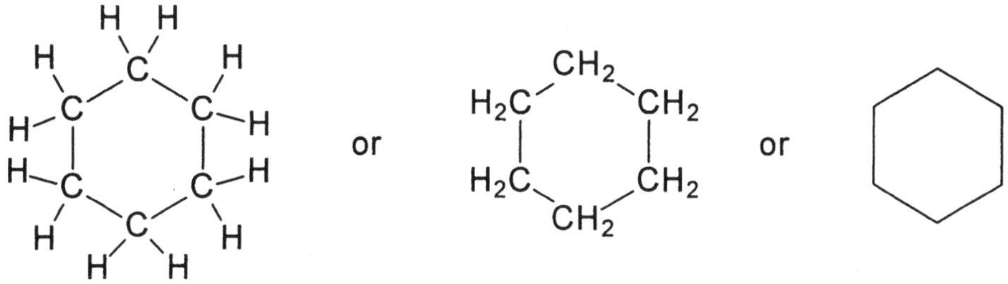

Note that cyclic and open chain structures with the same number of carbon atoms are *not* isomers of each other, since they do not have the same molecular formula.

Rules of Systematic Nomenclature: Alkanes

Systematic naming of the alkane hydrocarbons involves just three simple steps, shown here for the following seven carbon structure:

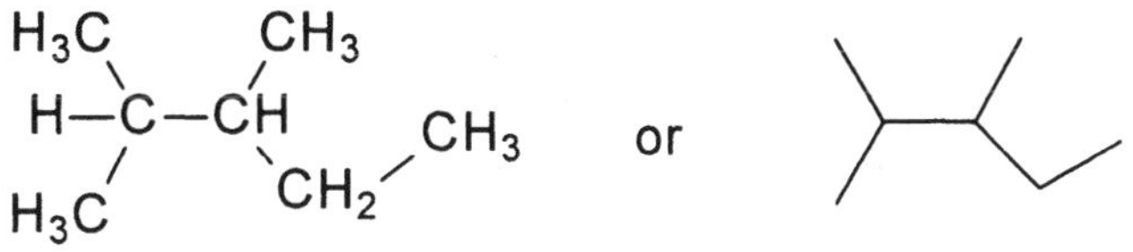

1. Identify the *longest continuous chain* of carbon atoms in the structural formula; the compound will be named as a derivative of that alkane, pentane, in this case.

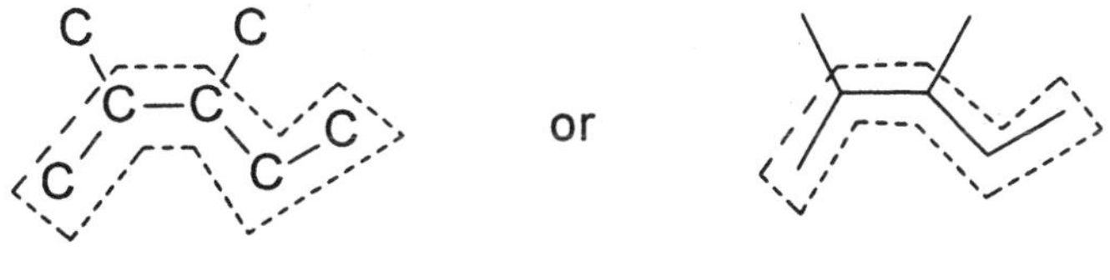

2. Number the chain to identify the position of each attached group, which in this case are both methyl (CH_3) groups. Start numbering the carbon atoms from either the left or right end of the longest chain so as to specify the positions of attached groups by using the *smallest* numbers. (In this case numbering must start from the left to comply with the smallest number rule.)

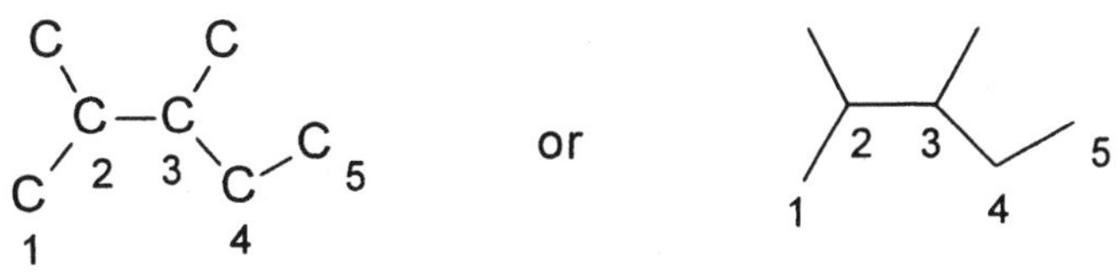

69

3. Identify each attached group by name. If more than one attached group of the same type exists, identify that fact by the prefix *di-*, *tri-*, *tetra-*, etc.

Thus the name of the compound in this example is *2,3-dimethylpentane*. Like any good systematic name, this unambiguously describes only one unique chemical structure.

Physical Properties of Alkanes

All the liquid alkane hydrocarbons have densities less than that of water and all are totally insoluble in water. Thus, mixing any of the liquid hydrocarbons with water will form an immiscible two-layer system, with the hydrocarbon layer floating on top. Like all homologous series of organic compounds, the paraffin hydrocarbons have boiling temperatures that increase in a regular fashion as the length of the carbon chain increases. Melting points also increase in a regular fashion with molecular size. The abbreviated table below lists the boiling temperatures of some of the normal (straight chain) alkanes that are liquids at room temperature. In fact, a plot of boiling point versus the number of carbon atoms in the chain produces a remarkably smooth curve in this and other homologous series of organic compounds, as shown in Fig. 12-2.

Compound	Boiling Point (bp) °C
n-pentane	36
n-hexane	69
n-heptane	98
n-octane	126
n-nonane	151
n-decane	174

Fig 18-2: Boiling temperatures of some alkane hydrocarbons.

Molecular Shape of the Alkanes

The optimal shape of saturated compounds is tetrahedral, or very nearly so. It is this shape that allows each of the four bonding electron pairs maximum spatial separation to minimize negative charge repulsion.

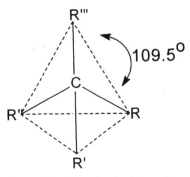

Fig 18-3: Tetrahedral bond angles in the alkanes.

In many *cycloalkanes,* however, it may not be geometrically possible to achieve these optimal bond angles. In cyclobutane and cyclopropane, angles would appear to be constrained to 90° and 60°, respectively, by purely geometric representation. (Actually these small ring compounds may come closer to tetrahedral *orbital shapes* at each carbon atom, at the cost of less orbital overlap with neighboring carbons. This results in weaker bonds between adjacent carbons.) This "angle strain" in the molecule decreases its stability and increases its stored energy content. Not surprisingly, such molecules tend to release a large amount of energy in chemical reactions that result in opening the ring.

Barriers to free-rotation in open chain alkanes — and other open chain compounds with only single bonds — is very small. As a result, all the C-C and C-H bonds in the alkanes rotate freely at all but the very coldest temperatures:

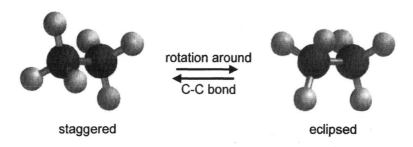

Small energy barriers between rotational forms do exist, however, giving rise to "favored" or most stable conformations:

Ball and Stick Models

rotation around

C-C bond

staggered   eclipsed

Newman Projections

60°
angles

0°
angles

staggered   eclipsed

Fig 18-4: Staggered and eclipsed conformations in ethane.

The staggered and eclipsed rotational forms of ethane are most clearly indicated in the *Newman Projections,* shown above. In this representation, the C-C bond is oriented perpendicular to the plane of the paper. The circle represents the front C atom, with its 3 attached hydrogen atoms; the rear carbon is in back of the front carbon and so is not visible, but its attached hydrogen atoms are. This representation clearly shows the staggered or eclipsed arrangement of the hydrogen atoms as C-C bond rotation occurs. The eclipsed form is *less* stable (contains more energy) due to the closer proximity of the electron pairs of the three C-H bonds on the front and back carbon atoms, with resulting electron pair repulsion. Eclipsing these bonds is said to produce *torsional strain* in the molecule.

The potential energy diagram below shows the increased energy of the eclipsed compared to the staggered forms as rotation proceeds:

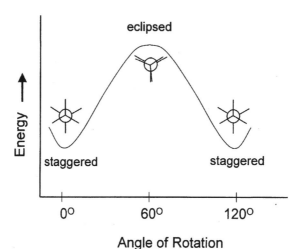

Fig 18-5: Rotation about the bonded carbon atoms in ethane.

The energy barrier in this case is low enough to allow rotation to freely occur at room temperature, but the dynamic equilibrium will greatly favor a higher percent composition of the more stable staggered form. Not surprisingly, as the groups attached to the central C-C bond increase in size, the energy barrier to free rotation gets larger. Thus, the eclipsed methyl-methyl overlap in butane contains 4.5 kcal/mole more energy than the staggered (*anti-*) conformation, compared to only a 2.8 kcal/mole difference between the eclipsed and staggered conformers of ethane.

Eclipsed conformation
of butane

Staggered (anti)
conformationof butane

Fig 18-6: Eclipsed and staggered (anti-) conformations of n-butane.

Cyclohexane

   Cyclohexane is a flexible molecule. The six carbon atoms are joined together to form a six-membered ring, but the molecule is capable of a limited amount of twisting, or *puckering,* to achieve the most stable possible tetrahedral bond angles (very close to 109.5°) and to minimize torsional strain. (Clearly the molecule will greatly disfavor a planar shape since a planar hexagon has internal bond angles of 120° by geometric law. A planar form would also have torsional eclipsing between all adjacent H atoms, further destabilizing this structure.)

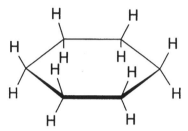

Fig 18-7: Planar cyclo-
hexane does not exist due to
angle and torsional strain.

   "Puckering" of the ring allows formation of the most stable "chair" conformation, in which both angle strain and torsional strain are virtually absent. (In a puckered conformation, three alternate carbon atoms exist *above* an imaginary central plane and the other three alternate carbon atoms exist *below* this plane.)

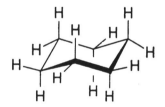

Fig 18-8: The "puckered"
chair conformation of cyclohexane. This most stable
conformation minimizes
angle and torsional strain.

   Various other conformations of cyclohexane are produced as the molecule twists from one chair form to its mirror image chair form. All these various intermediate forms have some degree of angle and torsional strain and thus are disfavored by the dynamic equilibrium that inter-converts these structures:

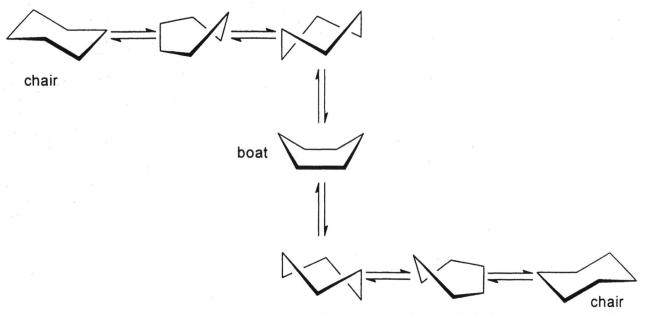

Fig 18-9: Various conformations of cyclohexane. The two mirror image chair forms are most stable.

Alkane Chemistry

Preparation of Alkanes

Alkanes can be prepared by the addition of hydrogen gas across the double bond of an *alkene*. In this re-action, which is very easy to visualize, the H-H bond is broken and the pi bond of the alkene breaks (opens) to receive the two hydrogen atoms:

The function of the *catalyst* (usually platinum, Pt, or palladium, Pd, deposited on microscopic parti-cles of carbon black, C) is to increase the ease of breaking bonds in the reacting molecules. This allows the reaction to proceed much faster than it would in the absence of the catalyst. A catalyst is defined as a chemical agent that enhances the rate of a reaction without itself being permanently changed. (We'll see many more examples of reaction catalysis as we continue learning organic chemistry.)

Alkanes can also be prepared from alkyl halides via Grignard reagents, by a method best described in a later section.

74

Reactions of Alkanes

Combustion

Alkanes are the least reactive of the organic molecules. They will, however, like any molecule containing carbon and hydrogen, react with oxygen to form CO_2 and water:

$$Alkane + O_2 \longrightarrow CO_2 + H_2O$$

This reaction, called combustion (burning), is very highly exothermic and liberates a great deal of energy. We utilize this energy to power our vehicles with gasoline or kerosene (jet fuel), and to heat our homes with natural gas or heating oil.

Halogenation

Substitution of halogen atoms F, Cl, or Br, for alkane hydrogen atoms occurs in the gas-phase under the influence of UV radiation or at very elevated temperatures. This free-radical chain reaction occurs as shown below. (See Chapter 21 for a detailed discussion of this reaction.)

$$RH + X_2 \xrightarrow[\text{light}]{\substack{\text{heat} \\ \text{or}}} RX + HX$$

alkane   halogen   alkyl halide   haloacid

The symbol R, above, is frequently used to represent a *generalized* alkyl structure. (The term *alkyl* means derived from an alkane; RH refers to the alkane itself.) This symbol is frequently employed in organic chemistry to indicate the *general* nature of a reaction or property, implying here that most any alkane or alkyl compound will undergo a given reaction or display a given property.

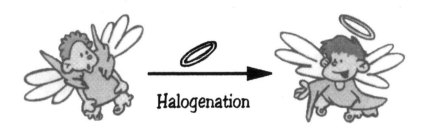

Halogenation

Chapter 19
Alkyl Halides: Substitution and Elimination Reactions

Alkyl halides are compounds having a saturated carbon atom directly bonded to a halogen atom, X. The bond is polarized as shown below because of the greater electronegativity of the halogen atom, which attracts electrons more strongly than the carbon atom. This results in a partial positive charge on the carbon and a partial negative charge on the halogen. This permanent bond polarization is basic to understanding the chemical reactivity of the alkyl halides.

$$\underset{\delta+ \quad \delta-}{>\!C\!-\!X}$$

dipole

Nucleophilic Substitution of Alkyl Halides

A *nucleophile* is an ion or atomic grouping bearing at least one electron pair that it can donate to form a covalent bond; nucleophiles are therefore Lewis bases. In the nucleophilic substitution reactions of alkyl halides, various nucleophiles replace the halogen to produce a wide range of products, including other alkyl halides, alcohols, ethers, amines, and products with new carbon-carbon bonds, among others:

$$Nuc^{\ominus} + {>}C{-}X \longrightarrow {>}C{-}Nuc + X^{\ominus}$$

| nucleophile | alkyl halide | | substitution product | halide ion |

Alkyl halide	Nucleophile	Substitution Product
RX	$^{\ominus}OH$, H_2O	ROH (Alcohol)
RX	$^{\ominus}OR'$, HOR'	ROR' (Ether)
RX	$^{\ominus}CN$	RCN (Nitrile)
RX	$^{\ominus}C{\equiv}CR'$	RC≡C-R' (Alkyne)
RX	$^{\ominus}X'$	RX' (Halide)
RX	$^{\ominus}O\text{-}\overset{O}{\underset{\parallel}{C}}\text{-}R'$	$R\text{-}O\text{-}\overset{O}{\underset{\parallel}{C}}\text{-}R'$ (Ester)
RX	$^{\ominus}NH_2$, NH_3	$R\text{-}NH_2$ (Amine)
RX	$^{\ominus}R'$	R-R' (Hydrocarbon)

Fig 19-1: Reactants and products in nucleophilic substitution reactions.

Nucleophilic substitution reactions of alkyl halides can occur by either of two mechanisms, both of which depend on the polarization of the carbon-halogen bond.

1. The Concerted, One-Step, S_N2 Mechanism

By this route the nucleophile attacks and forms a bond to the backside of the electron deficient carbon, with the halide leaving *simultaneously* from the front side. The driving force for this reaction is the attractive force between opposite charges—the negative electron pair of the nucleophile and the (partial) positive charge on the polarized carbon atom.

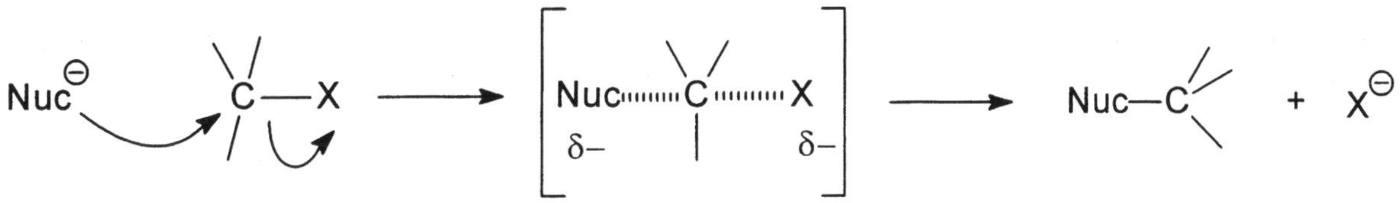

Fig 19-2: The one-step S_N2 mechanism.

This reaction mechanism is called S_N2, which stands for Substitution, Nucleophilic, second order. The first terms are self-explanatory; second order refers to the fact that the rate of a reaction proceeding by this mechanism depends on the concentration of *both* (two) reacting species.

2. The Two-Step S_N1 Reaction Mechanism

In this mechanism, ionization of the C-X bond occurs in a slow first step. This heterolytic bond cleavage releases the halide, X⁻, and produces a carbocation intermediate. The carbocation then reacts rapidly with the nucleophile in a second step to produce the substitution product:

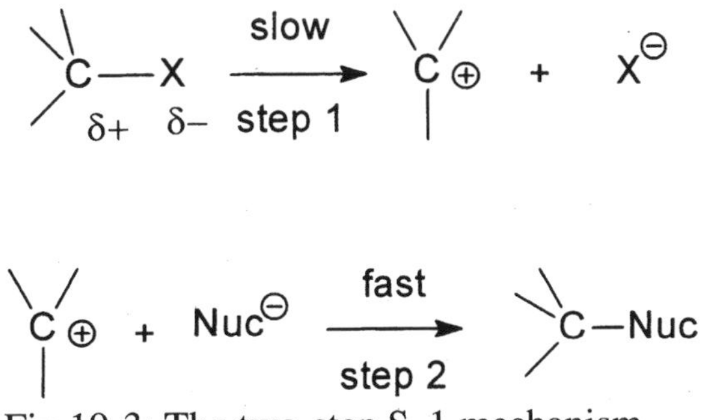

Fig 19-3: The two-step S_N1 mechanism.

The first step, formation of the carbocation, is the *slow* step, which limits and controls the rate of the overall reaction. Here, the reaction rate depends on the alkyl halide concentration since it is the precursor of the slowly forming carbocation. Reaction rate does not depend on the nucleophile concentration, since it is involved only in the fast second step. Since the rate depends only on the concentration of *one* species, the alkyl halide, this mechanism is called S_N1, an abbreviation for: Substitution, Nucleophilic, first order.

The rate of the reaction by *either* mechanism is aided by the polarization of the C-X bond. Bond polarization is a necessary condition for this type of reaction to occur. The greater the partial positive charge on the carbon atom, the more attractive that site is for S_N2 attack by the electron-rich nucleophile.

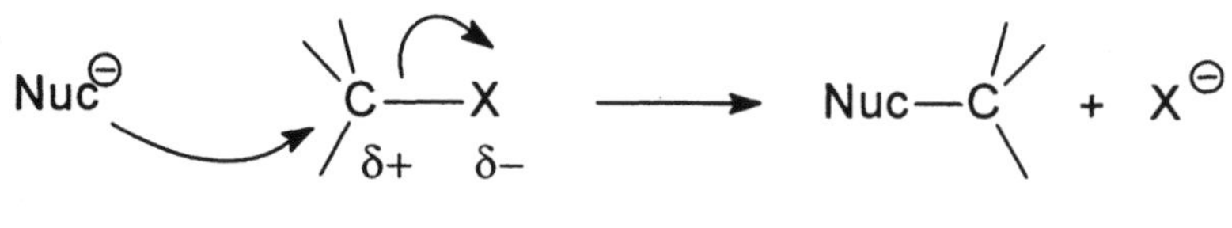

Bond polarization also eases and encourages the formation of the carbocation by the S_N1 mechanism, since polarization causes the bond to have a degree of charge separation even before reaction commences:

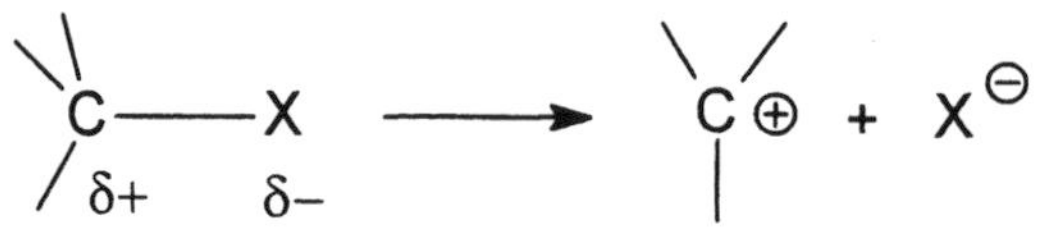

partial charge   full charge
separation   separation

Factors Favoring S_N1 and S_N2 Mechanisms

One of the two substitution mechanisms generally predominates over the other. Which mechanism predominates depends on the details of the specific reaction. A few simple considerations help to predict which mechanism will be favored in a given reaction.

Bond Polarity

High polarity of the C-X bond will favor the rate of reaction by either mechanism.

Alkyl Halide Structure

S_N1 reactions are favored by factors that result in greater carbocation stability. *The more carbon atoms which are bonded to the carbon bearing the positive charge, the more stable is the carbocation and the faster it forms.* Tertiary carbocations, which form with the greatest ease, are more stable than secondary carbocations. Primary, and especially methyl, carbocations are the least stable and form most slowly and with the greatest difficulty, if they can form at all. The terms primary (1°), secondary (2°), and tertiary (3°) refer to the number of carbon atoms bonded to the carbon bearing the positive charge,1,2, and 3, respectively. Primary, secondary, and tertiary carbocations result from the ionization of primary, secondary, and tertiary alkyl halides, which have, respectively, 1, 2, or 3 carbons bonded to the carbon linked to the halogen.

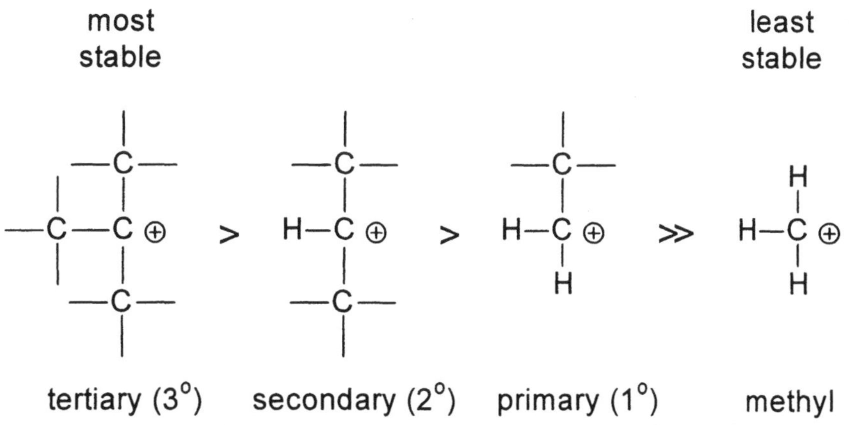

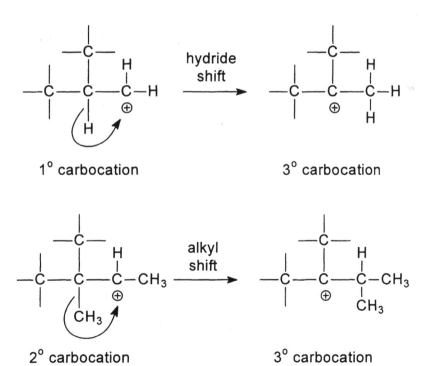

Fig 19-4: Order of carbocation stability and ease of formation.

Thus, tertiary halides tend to react by an S_N1 (carbocation) mechanism, while primary and methyl halides generally react by a one-step, S_N2 mechanism. Secondary halides can react by either or both mechanisms. *Carbocation stability is probably the most important factor determining which of these mechanisms will prevail for a given reaction.*

An initially produced carbocation will rapidly rearrange to a more stable carbocation, if it can do so by a hydride (H^-) or alkyl (R^-) shift from an *adjacent* carbon atom. For example:

Fig 19-5: Rearrangement of carbocations by hydride or alkyl shift.

This has the important practical consequence that the isolated substitution product is often the isomer derived from the rearranged carbocation, not the initially produced carbocation. For example:

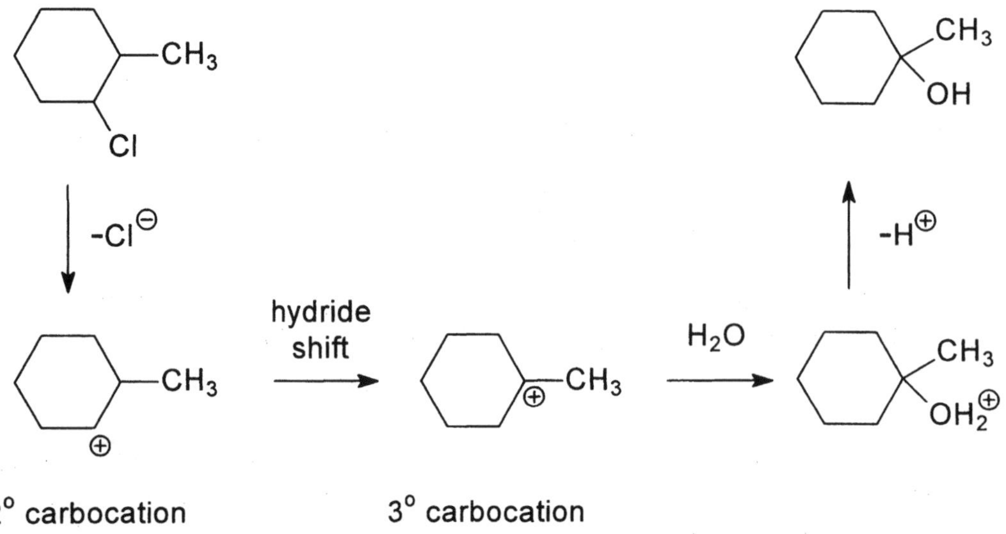

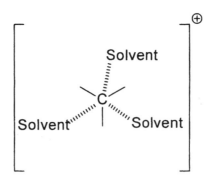

2° carbocation   3° carbocation

Fig 19-6: Formation of a 3° alcohol resulting from rearrangement of a carbocation intermediate.

Solvent Effects

Another factor that stabilizes and encourages the formation of carbocations is *high solvent polarity.* If the substitution reaction occurs in a highly polar solvent, the carbocation is stabilized by interaction with solvent molecules. This interaction disperses some of the charge over several solvent molecules as well as the positive carbon atom, and thus favors an S_N1 mechanism. The carbocation mechanism is especially favored in *protic* solvents such as alcohol or water; since these are especially good at solvating and stabilizing ions. Polar, *aprotic* solvents also encourage S_N2 chemistry, since these are polar reactions as well. Such solvents include DMF (dimethylformamide), DMSO (dimethylsulfoxide), and DMA (dimethylacetamide). Substitution reactions by either mechanism are especially slow in non-polar solvents.

Fig 19-7: Carbocation stabilization by a polar solvent.

Steric Hindrance

Factors that increase the difficulty of approach by the attacking nucleophile make S_N2 reactions more difficult, since the nucleophile is inhibited from easy access to the backside of the carbon bearing the halogen atom. This is an example of *steric hindrance,* where proximity of a bulky atom or atomic group blocking the site of a potential reaction slows down the rate of that reaction. Steric hindrance at the site of substitution thus favors a carbocation, S_N1, mechanism, since close approach of the nucleophile to the hindered carbon does not occur in the slow step of that mechanism.

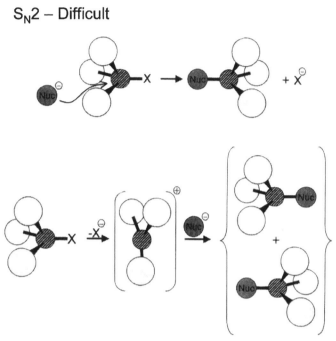

Fig 19-8: Steric hindrance in nucleophilic substitution.

Nucleophile Concentration

High nucleophile concentration favors S_N2 reaction, since the rate of these reactions depends directly on the concentration of the attacking nucleophile, while the S_N1 reaction does not. This is easy to visualize, since the collision rate between the reacting molecules increases with the concentration of both. A high collision rate favors the formation of the transition state needed for the S_N2 reaction to proceed. By contrast, the S_N1 reaction is unaffected by nucleophile concentration. The rate of S_N1 reactions is limited by the slow rate of formation of the carbocation, which depends on the concentration of *only* the alkyl halide.

Strength of the Nucleophile

Some nucleophiles react avidly; others are relatively unreactive. Strong nucleophiles react aggressively to attack the positively polarized carbon of the alkyl halide. S_N2 reactions proceed rapidly in reactions of strong nucleophiles. By contrast, the rates of S_N1 reactions would be unaffected by nucleophile strength.

SUMMARY

The Two-Step S_N1 reaction mechanism is favored for tertiary (3°) alkyl halides, which can produce stable tertiary carbocations; by a highly polar protic solvent; and by steric hindrance afforded by bulky groups blocking the site of nucleophile attack by a competing S_N2 mechanism.

The One-Step S_N2 reaction mechanism is favored by primary (1°) alkyl halides, which offer minimum steric hindrance to an attacking nucleophile. In addition, primary alkyl halides or methyl halides produce unstable primary or methyl carbocations, thus retarding the competing S_N1 reaction; S_N2 reactions are also favored by polar solvents; by high nucleophile concentrations; and by highly reactive nucleophiles.

Elimination Reactions of Alkyl Halides

Alkyl halides, which possess at least one neighboring hydrogen atom on a saturated carbon, can *eliminate* the elements of the acid HX to form alkenes:

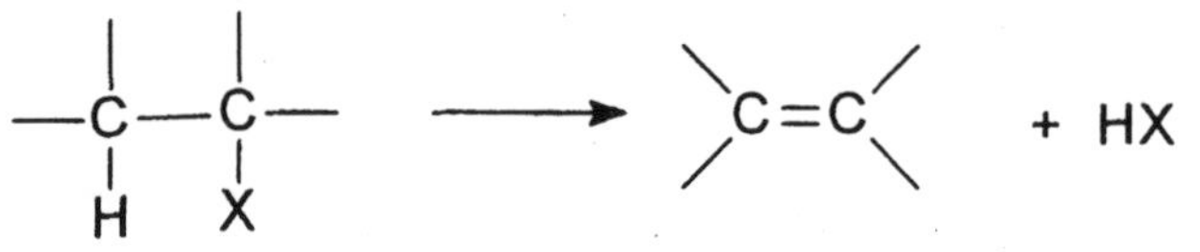

Like the substitution reactions described in the preceding section, these elimination reactions can also occur by either of two mechanisms, described as E1 and E2. These designations stand for elimination-first order and elimination-second order, respectively. The terms first and second order refer to the dependence of reaction rate on either one or two reacting species. The E1 mechanism proceeds, like S_N1, through a carbocation intermediate:

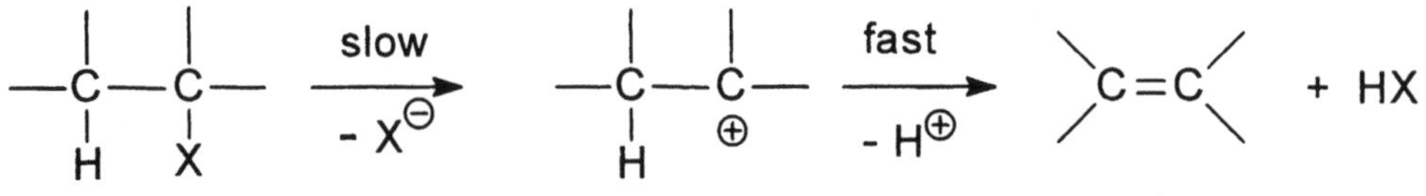

Fig 19-9: Elimination by a two-step E1 mechanism.

The slow step (rate-determining step) of this reaction is the formation of the unstable carbocation. Since this rate depends *only* on the concentration of the starting alkyl halide, this is called an E1 mechanism.

In the presence of a suitable base, the loss of hydrogen (as a proton) and the neighboring halogen (as halide) can also occur *simultaneously* in a single concerted step. The rate of reaction by this mechanism depends on the collision frequency between the base and the alkyl halide and thus depends on the concentration of both. This mechanism is referred to as E2 for that reason.

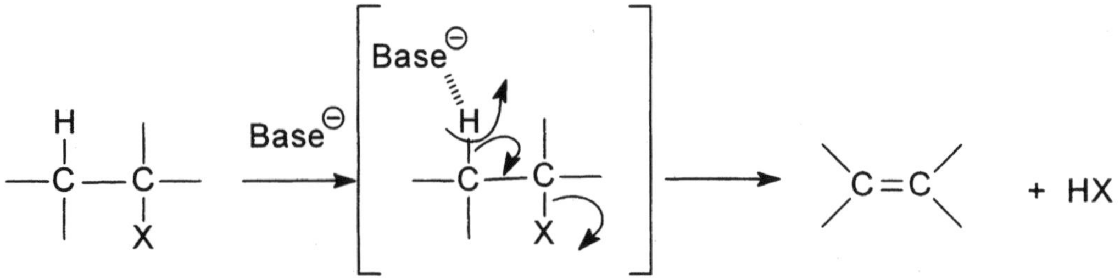

Fig 19-10: Elimination by a one-step E2 mechanism.

Substitution and elimination reactions often occur in competition with each other and can occur by both carbocation and concerted mechanistic pathways:

Substitution (S$_N$1) / Elimination (E1)
via Carbocation Mechanism

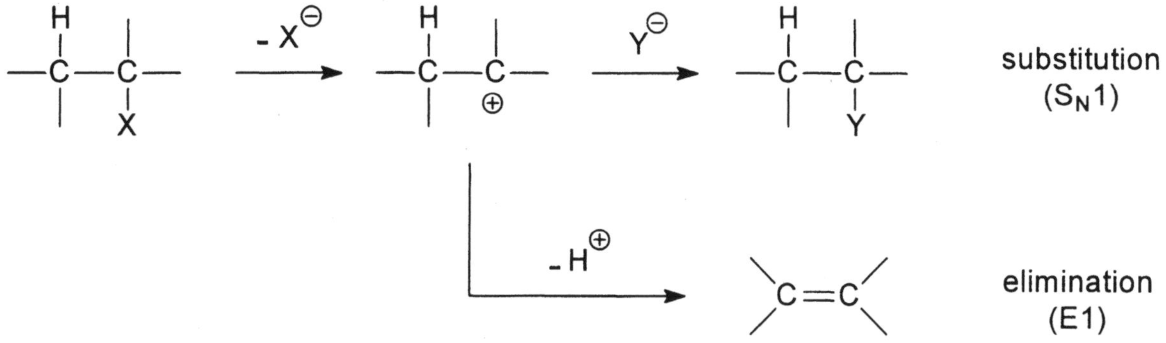

substitution
(S$_N$1)

elimination
(E1)

Substitution (S$_N$2) / Elimination (E2)
via Concerted Mechanism

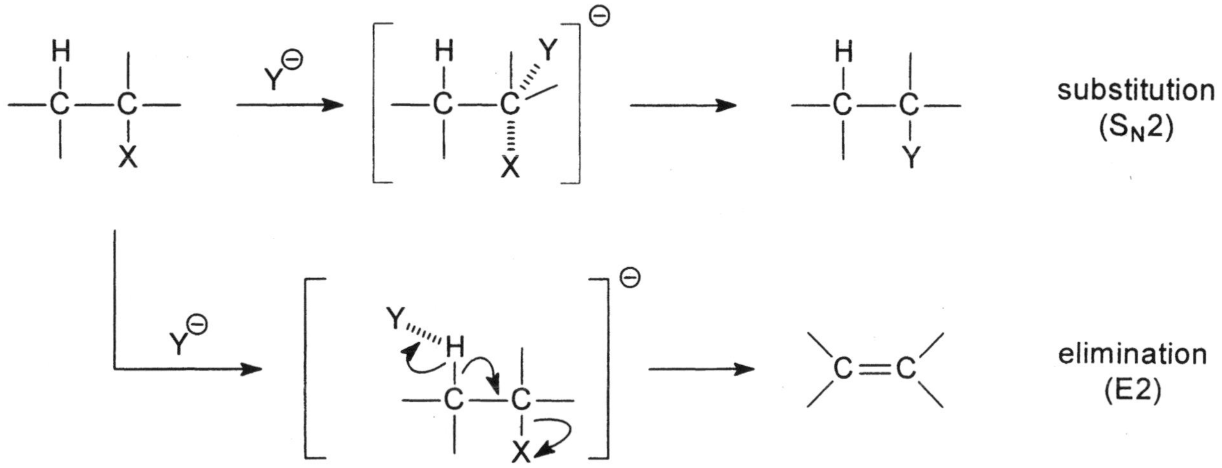

substitution
(S$_N$2)

elimination
(E2)

Fig 19-11: Summary of substitution and elimination mechanisms. (In this scheme Y$^-$ can serve either as a substituting nucleophile in the S$_N$2 reaction or as a base in the E2 reaction.)

Chapter 20
The Unsaturated Hydrocarbons: Alkenes and Alkynes

Alkenes, also called *olefins,* possess a double bond linking two carbon atoms. One of these is the sigma bond directed along a line joining the two atoms. The other is the pi bond resulting from the overlap of the dumb-bell shaped p orbitals above and below the bond axis. The sigma bonds are sp^2 hybrids, and thus alkenes are planar with bond angles of 120 degrees as predicted by the VSEPR theory for this AX_3 system:

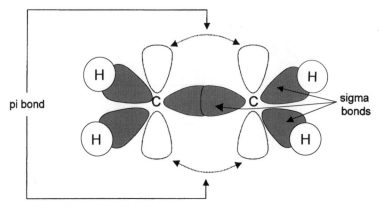

Fig 20-1: Bonding in alkenes.

The double bond is represented by a double line joining the two carbon atoms, as shown below for ethylene, the simplest of the alkenes, and several examples of other alkenes:

ethylene   cyclohexene   1,3-butadiene

$CH_3(CH_2)_7CH=CH(CH_2)_7COOH$

oleic acid   methyl methacrylate

vitamin A

Reactions of the Alkenes

Most of the important reactions of alkenes may be classified into one of two categories, with some overlap between them. These two catagories are *addition* and *oxidation* reactions.

Addition Reactions

The alkene (olefinic) pi bond can open to accept other atoms or atomic groupings to form saturated addition products. In general:

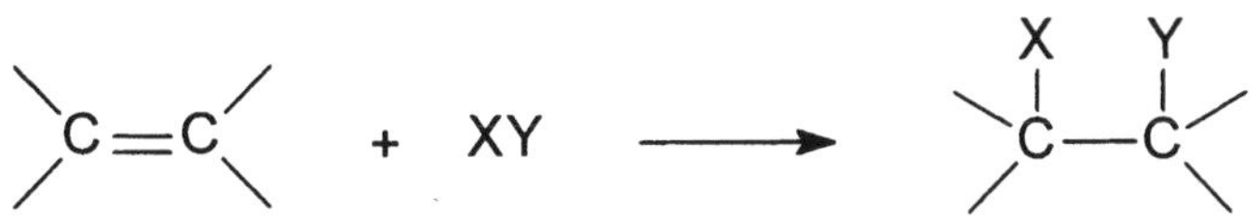

Catalytic Hydrogenation

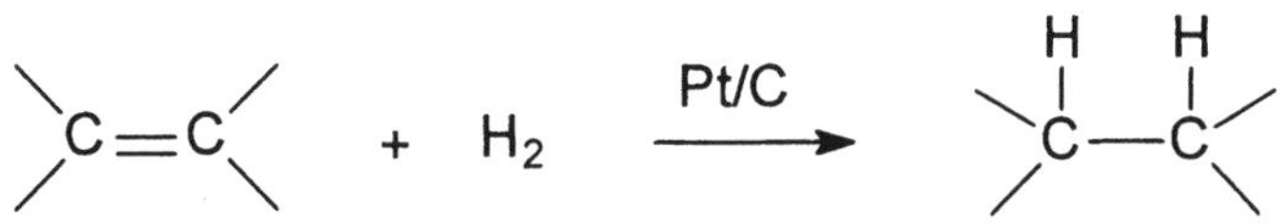

alkane

Addition of Halogen Acids

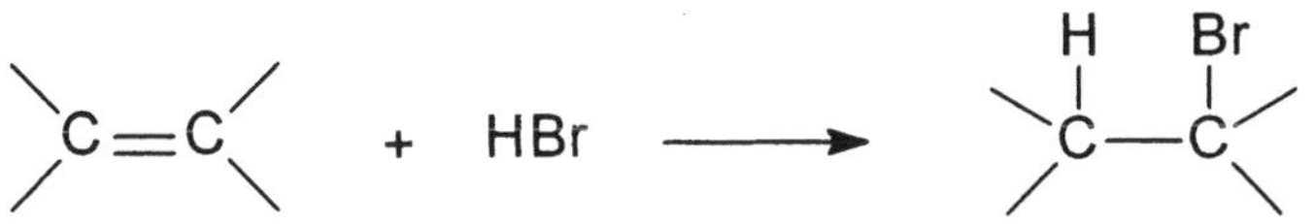

alkyl halide

Addition of Molecular Halogens

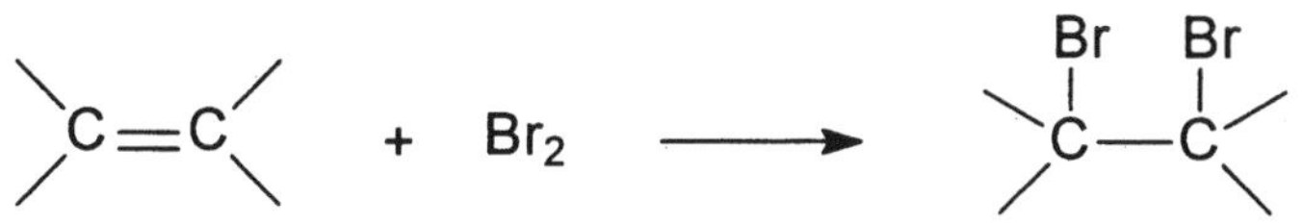

dihalide

Addition of Water (Hydration)

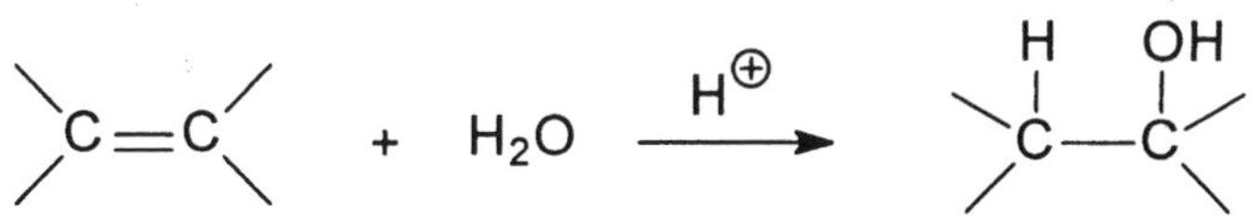

alcohol

Halohydrin Formation

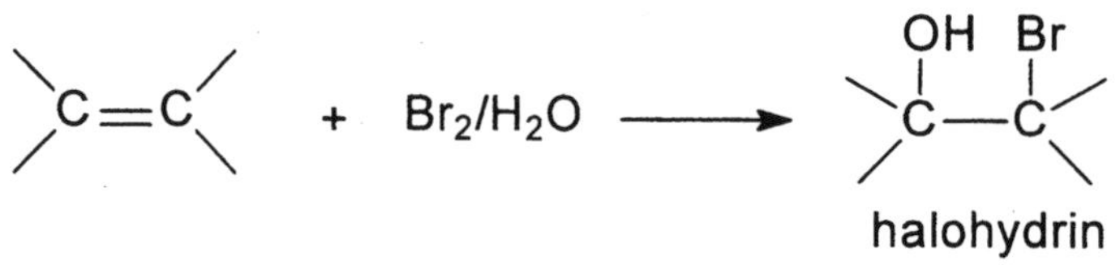

$$\text{C}=\text{C} \quad + \quad Br_2/H_2O \quad \longrightarrow \quad \overset{OH \quad Br}{\text{C}-\text{C}}$$

halohydrin

Oxidation Reactions

The second important category of alkene reactions is oxidation. These oxidations can be broken down into two subclasses.

I. Addition/ Insertion Oxidations
   Cis -hydroxylation

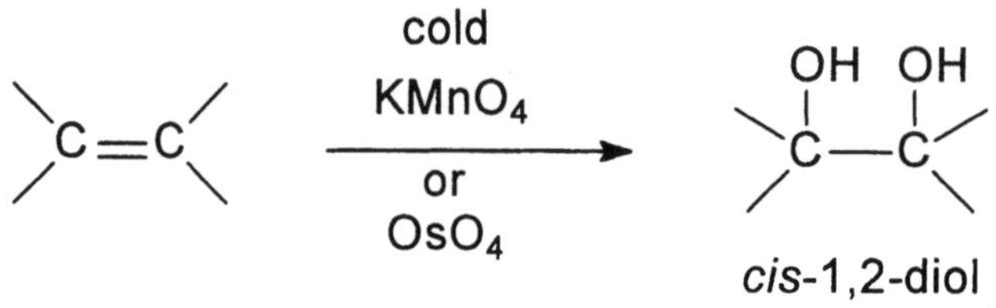

$$\text{C}=\text{C} \quad \xrightarrow[\substack{or \\ OsO_4}]{\substack{cold \\ KMnO_4}} \quad \overset{OH \quad OH}{\text{C}-\text{C}}$$

cis-1,2-diol

Epoxidation

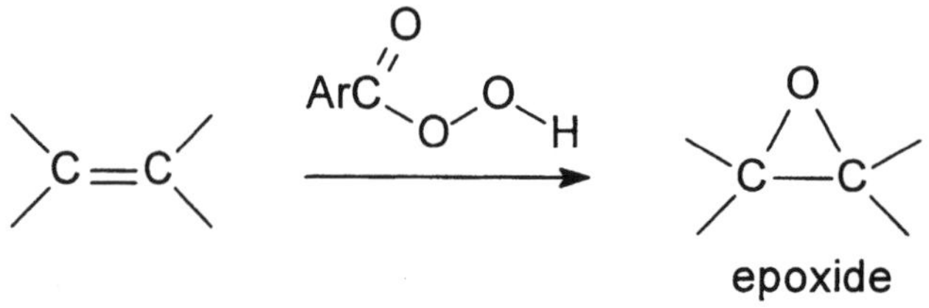

$$\text{C}=\text{C} \quad \xrightarrow{ArC(=O)O-O-H} \quad \overset{O}{\text{C}-\text{C}}$$

epoxide

2. Oxidative Cleavage Reactions
   In these reactions, cleavage (fragmentation) occurs between the two alkene carbon atoms, indicated by the wiggly line. (Details of these reactions follow in the text.)
   Permanganate Oxidation

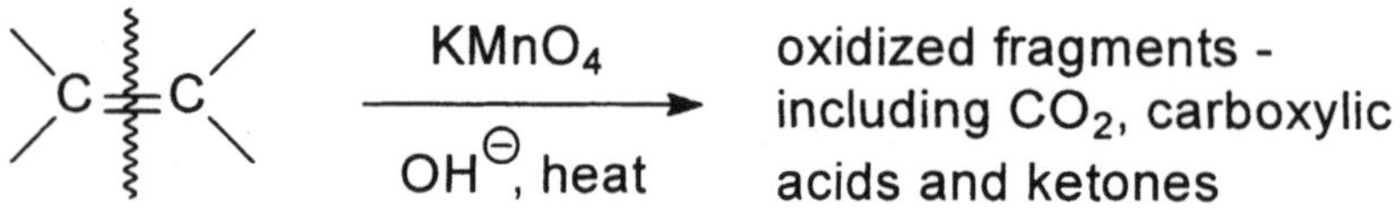

$$\text{C}\text{≀}\text{C} \quad \xrightarrow[\substack{OH^{\ominus}, \text{ heat}}]{KMnO_4} \quad \begin{array}{l}\text{oxidized fragments -} \\ \text{including } CO_2, \text{ carboxylic} \\ \text{acids and ketones}\end{array}$$

Ozonolysis

$$\text{C}\text{≀}\text{C} \quad \xrightarrow[\substack{2) H_2O, \text{ Zn}}]{1) O_3} \quad \begin{array}{l}\text{oxidized fragments -} \\ \text{including aldehydes} \\ \text{and ketones}\end{array}$$

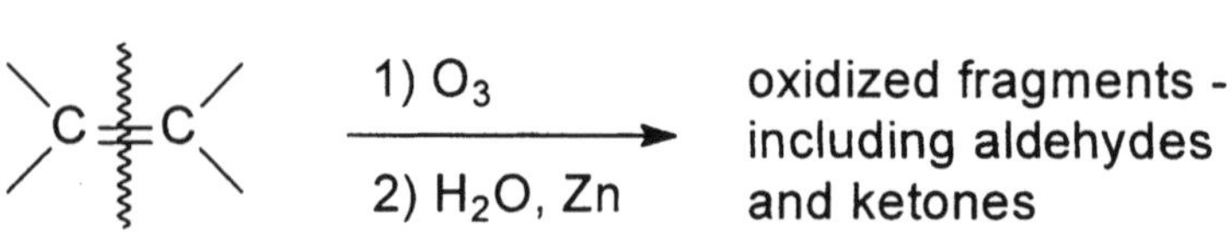

Alkene Reaction Mechanisms

Carbocation Addition Reactions

Alkenes generally behave as Lewis bases (electron pair donors) when reacted with mineral acids. First, the proton, H^+, adds to the pi electron pair of an olefin, forming a carbocation intermediate. This then combines with an electron donating species in the environment to complete the addition reaction and form a stable compound. The addition reactions of water and hydrogen bromide to alkenes are two important examples of this reaction. The Br^+ ion, derived from a molecule of Bromine, Br_2, can also add to alkenes forming a "bromonium" ion, which then reacts either with a Br^- ion or with water to complete the formation of a vic-dibromide or a halohydrin. (vic- means vicinal, referring to bonding to *adjacent* carbon atoms).

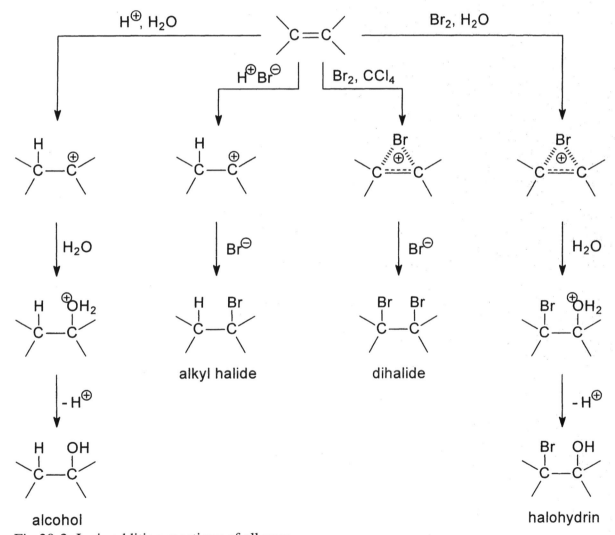

Fig 20-2: Ionic addition reactions of alkenes.

Addition To Unsymmetrical Alkenes

When *unsymmetrical* alkenes are reacted as in Fig 20-3, the predominant isomeric addition product is that derived from the *more stable* of the two possible carbocation intermediates:

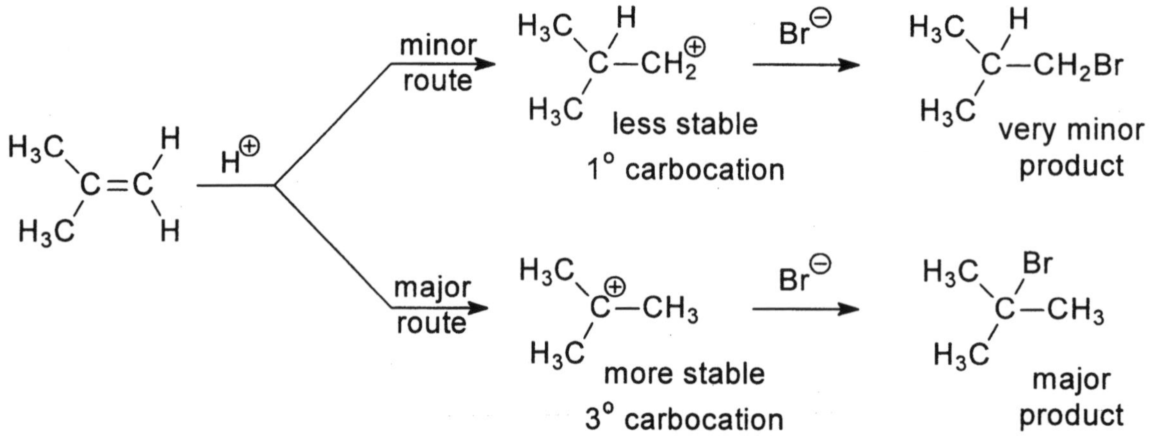

Fig 20-3: Major and minor products derived from 3° (more stable) and 1° (less stable) carbocations, respectively.

Catalytic Hydrogenation

This important reaction involves the addition of the two hydrogen atoms derived from a molecule of H_2 to the alkene. A heavy metal platinum or palladium catalyst is required for this reaction. Most often the metal is deposited on carbon black and the suspension is mixed with a solution of the alkene under an atmosphere of H_2 gas. The catalyst helps to break both the H-H bond and also the alkene pi bond. Both bond breaking events must occur for this reaction to proceed. The entire reaction, including bond-breaking and bond-forming steps, is believed to occur in close association with the surface of the catalyst:

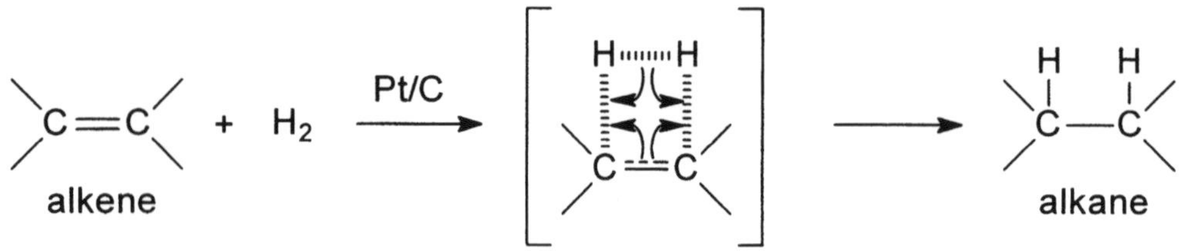

Oxidation Reactions

Addition

Cis- hydroxylation of alkenes produces a 1,2- diol, also called a vicinal (*vic*) diol. This is accomplished using either cold permanganate, MnO_4^- or osmium tetroxide, OsO_4, as the oxidizing agent. The reaction proceeds through an intermediate manganate or osmate ester which inserts both oxygen atoms into the double bond simultaneously from the same side to yield the *cis*-diol oxidation product:

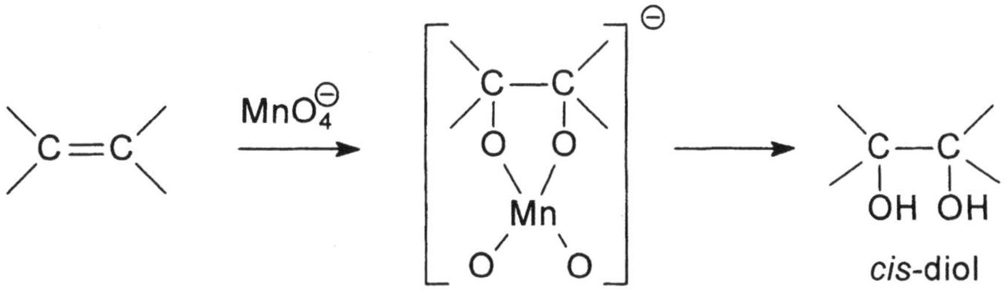

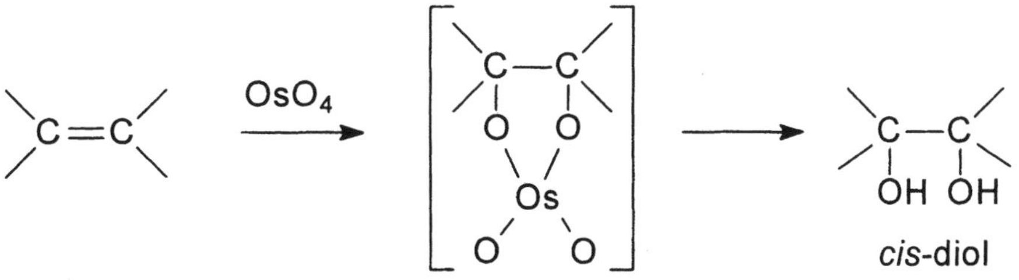

Oxygen Insertion

Peracids insert an oxygen atom into the alkene bond to produce an *epoxide*. The reagent used is most often MMPP (*mono-magnesium peroxyphthalate*). Epoxides, also called *oxiranes*, are valuable synthetic intermediates.

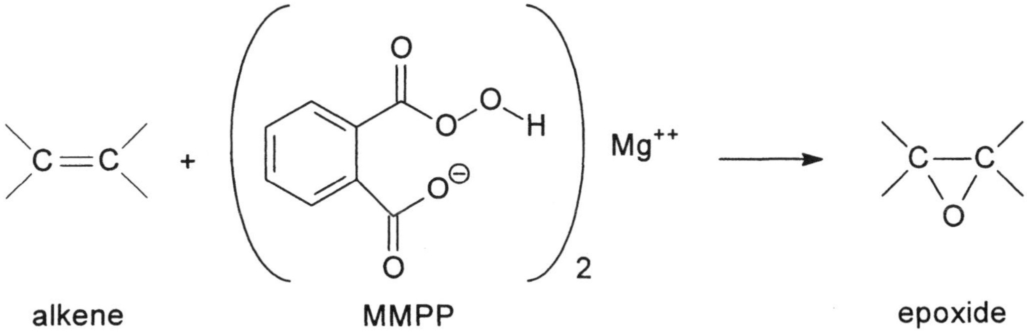

Oxidative Cleavage

Two important reactions of alkenes result in cleavage of the double bond to produce oxidized fragments. These methods employ hot potassium permanganate, $KMnO_4$, or ozone, O_3, as oxidizing agents.

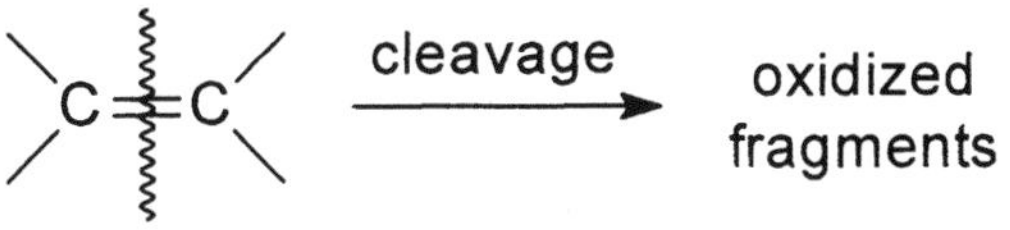

The oxidized fragments formed from a particular alkene structure can be predicted by applying a few simple rules. Indeed, by characterizing the structure of the oxidation fragments it is often possible to work backwards to establish the structure of the alkene. This was an important method for alkene structure determination prior to the advent of modern instrumental methods. The mechanistic details of these reactions need not concern us at this time, but the cleavage rules and products should be memorized.

OXIDATIVE CLEAVAGE RULES

Hot Permanganate Oxidation

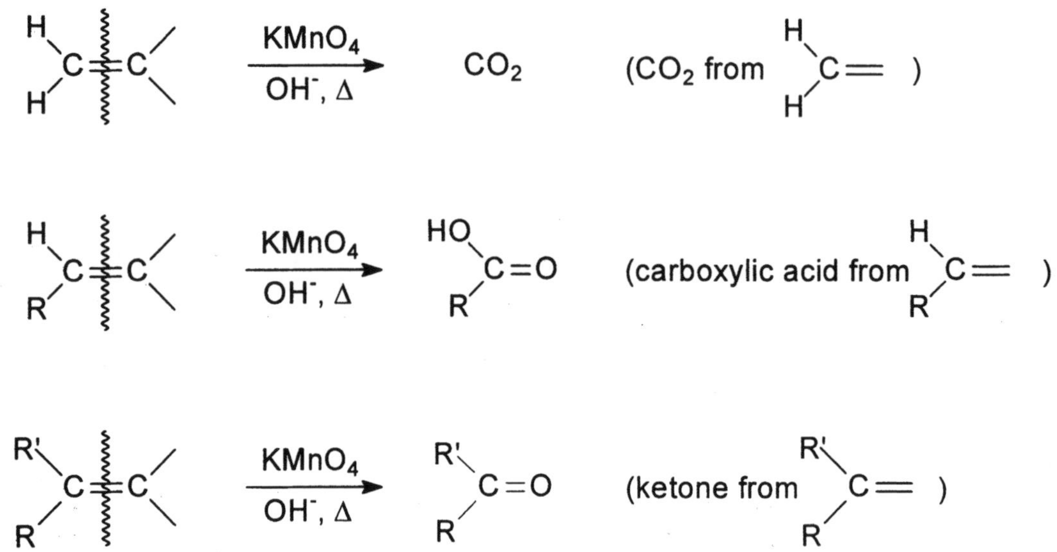

Fig 20-4: Oxidative cleavage of alkenes by hot permanganate.

Ozone Oxidation

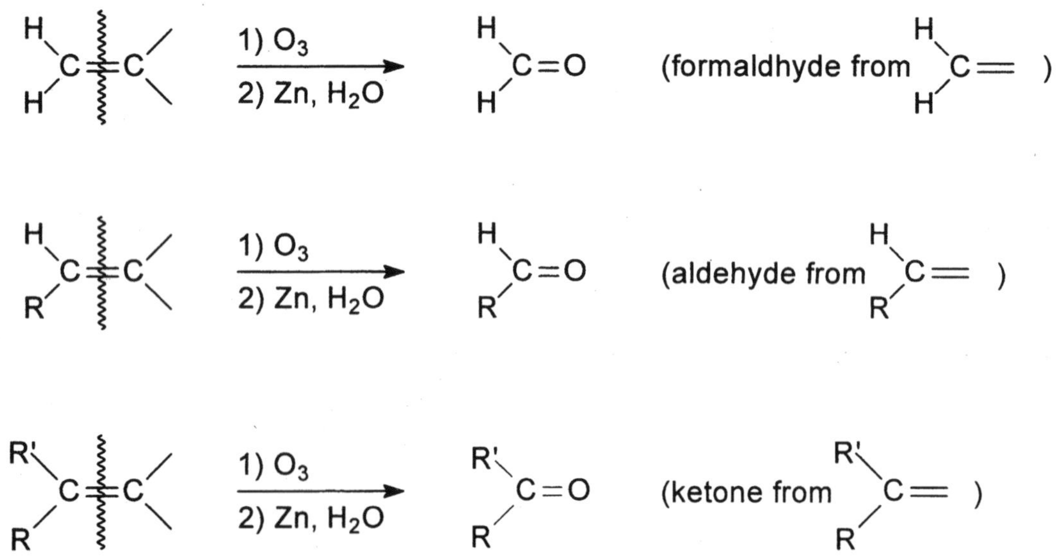

Fig 20-5: Oxidative cleavage of alkenes by ozone.

Here are a few specific examples of alkene oxidation:

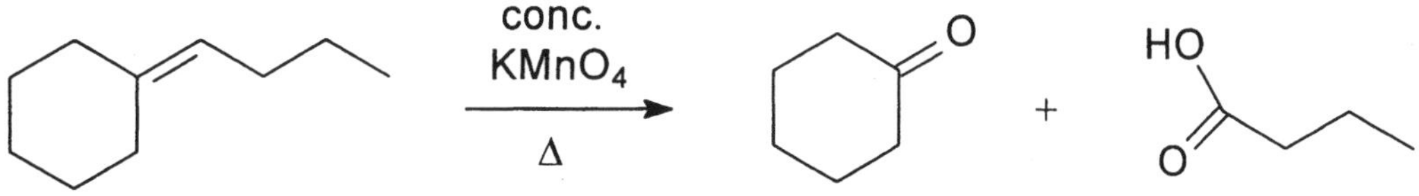

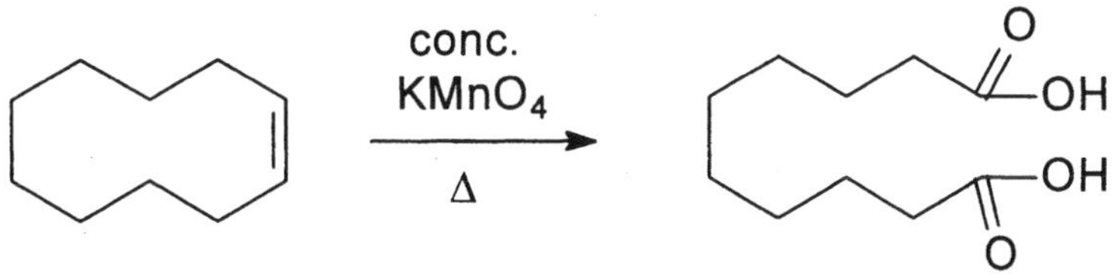

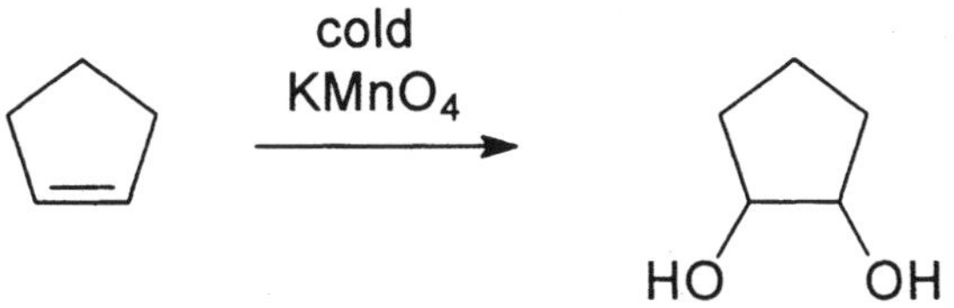

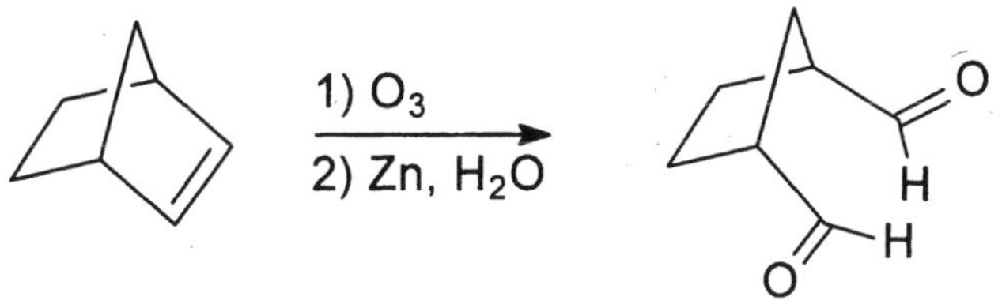

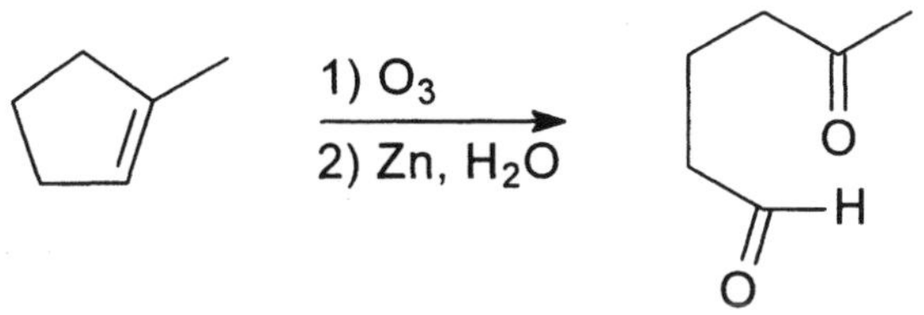

Reactions of Alkynes

Alkynes, having carbon-to-carbon *triple* bonds, can undergo much the same type of addition and oxidation reactions as alkenes, although with some easily understood differences. Each mole of an alkyne can add *two* moles of various reagents, since they possess two pi bonds, while alkenes only have one. Thus it is possible to add *two* moles of Br_2, halogen acids, H_2, etc., to alkynes. Oxidative cleavage reactions of alkynes produce fragments in a higher oxidation state than alkenes. The products from alkynes are carboxylic acids using either hot permanganate or ozone. One more interesting difference is that hydration (addition of water) of alkynes (a reaction catalyzed by Hg^{+2}), initially produces a *vinyl alcohol*. (The term *vinyl* refers to a group bonded directly to an alkene carbon atom.) Most vinyl alcohols spontaneously rearrange to a ketone; thus the isolated hydration product of an alkyne is a ketone. This isomerization is called

91

keto-enol tautomerization, and is a consequence of the greater stability of ketones compared to vinyl alcohols, in most cases. By contrast, the hydration of alkenes, of course, produces simple saturated alcohols.

$$R-C\equiv C-R' \xrightarrow[\text{Hg}^{++}]{\text{H}_2\text{O, H}^{\oplus}} \left[\begin{array}{c} R-C=C-R' \\ | \quad\; | \\ OH \;\; H \end{array} \right] \rightleftharpoons \begin{array}{c} R-C-CH_2R' \\ || \\ O \end{array}$$

Fig 20-6: Hydration of an alkyne followed by keto-enol tautomerization to form a ketone.

The use of certain specific hydrogenation methods allows for the stereospecific production of *cis-* or *trans-* alkenes by mono-hydrogenation of an alkyne. The Lindlar catalyst (Pd on CaCO$_3$) favors production of the cis alkene, while a dissolving metal reduction, using sodium in liquid ammonia and a proton source like ethyl alcohol, leads to the *trans* alkene.

A few examples of the important addition and oxidation reactions of alkynes follow. Each is a specific example of the general reactions described above.

$$H-C\equiv C-CH_2CH_3 + 2\ H_2 \xrightarrow{\text{Pt}} CH_3CH_2CH_2CH_3$$

$$CH_3-C\equiv C-CH_2CH_2CH_3 + H_2 \xrightarrow[\text{catalyst}]{\text{Lindlar}} \begin{array}{c} H \qquad\quad H \\ \diagdown C=C \diagup \\ CH_3 \diagup \qquad \diagdown CH_2CH_2CH_3 \end{array}$$

cis-alkene

$$CH_3-C\equiv C-CH_2CH_2CH_3 \xrightarrow{\text{Na / NH}_3} \begin{array}{c} CH_3 \qquad\quad H \\ \diagdown C=C \diagup \\ H \diagup \qquad \diagdown CH_2CH_2CH_3 \end{array}$$

trans-alkene

$$H-C\equiv C-CH_2CH_2CH_3 \xrightarrow{\text{2 HBr}} \begin{array}{c} Br \\ | \\ CH_3\overset{}{C}CH_2CH_2CH_3 \\ | \\ Br \end{array}$$

$$CH_3-C\equiv C-CH_2CH_3 \xrightarrow[\text{OH}^-,\ \Delta]{\text{KMnO}_4} \begin{array}{c} O \\ \diagup\diagup \\ CH_3-C \\ \diagdown OH \end{array} + \begin{array}{c} O \\ \diagdown\diagdown \\ C-CH_2CH_3 \\ HO \diagup \end{array}$$

$$CH_3-C\equiv C-CH_2CH_3 \xrightarrow[\text{2) Zn, H}_2\text{O}]{\text{1) O}_3} \begin{array}{c} O \\ \diagup\diagup \\ CH_3-C \\ \diagdown OH \end{array} + \begin{array}{c} O \\ \diagdown\diagdown \\ C-CH_2CH_3 \\ HO \diagup \end{array}$$

Chapter 21
Free Radical Reactions

The *homolytic* cleavage of a covalent chemical bond produces neutral free radicals. These radicals generally possess only *seven* electrons around each atom at the site of cleavage. They are unstable for this reason and eventually undergo chemical reactions to re-acquire the stable noble gas configuration of eight valence electrons.

$$A - B \longrightarrow A\cdot \ + \ \cdot B$$
$$\text{Free Radicals}$$

(The single dot represents the unpaired electron at the site of bond scission.) Free radicals, represented below as R•, react in only four important ways:

1. Addition to unsaturated systems:

$$R\cdot \ + \ \underset{/}{\overset{\backslash}{C}} = \underset{\backslash}{\overset{/}{C}} \longrightarrow R - \overset{|}{\underset{|}{C}} - \overset{/}{\underset{\backslash}{C}}\cdot$$

2. Combination with oxygen:

$$R\cdot \ + \ O_2 \longrightarrow R - O - O\cdot$$

3. Abstraction of H atoms:

$$R\cdot \ + \ H - \overset{/}{\underset{\backslash}{C}} \longrightarrow RH \ + \cdot\overset{/}{\underset{\backslash}{C}}$$

4. Covalent bond formation by combining with other radicals:

$$R\cdot \ + \ R'\cdot \longrightarrow R - R'$$

Note that (1), (2), and (3) above produce new radicals which will *continue to react* further; only (4) produces a covalent bond eliminating the unpaired electrons and forming a stable product with electron octets.

Decomposition of Benzoyl Peroxide

The formation of free radicals is favored by three factors: weak bonds, which are easily cleaved; high temperatures or energetic UV radiation to effect bond breakage; and the formation of stable products along with the unstable radicals. The decomposition of benzoyl peroxide to yield two moles of CO_2 and two phenyl radicals displays all of these factors:

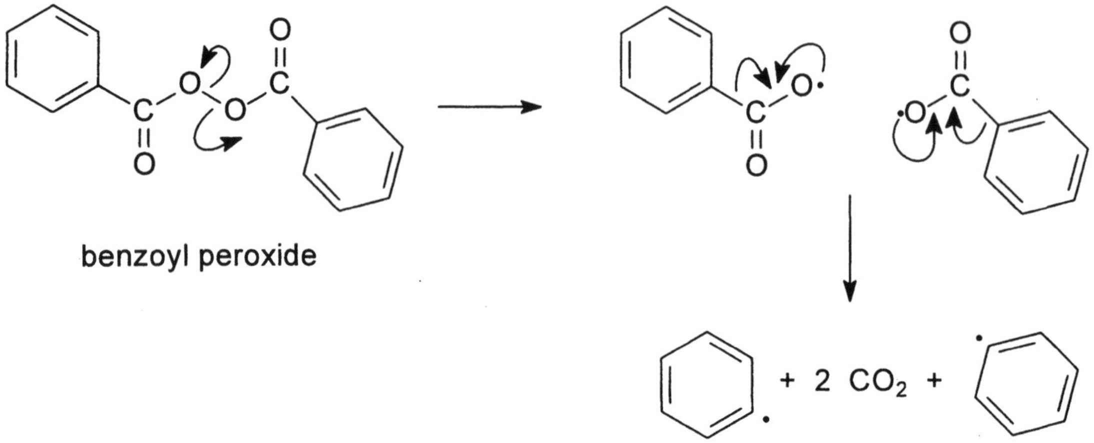

benzoyl peroxide

The ease of formation of radicals follows their order of stability, which is the same as for carbocations. Thus, radical stability follows the order $3° > 2° > 1° >$ methyl. As with carbocations, the most stable radicals form fastest and give rise to the highest yield of isolated products.

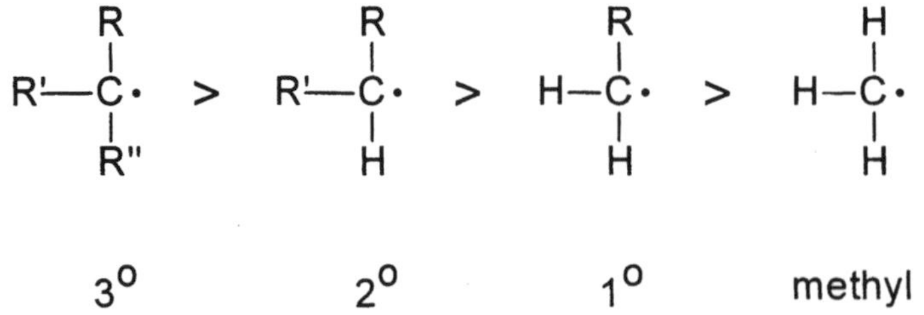

Fig 21-1: The order of stability of alkyl free radicals.

(Radicals in which the unpaired electron can be delocalized by resonance interaction with neighboring p orbitals have special stability and form much more easily, again following the same pattern as carbocations. See Chapter 27)

A few of the more important reactions and mechanisms involving free radicals are shown below.

Halogenation of Alkanes

Halogen atoms will substitute for hydrogen atoms in alkanes by a free-radical mechanism:

$$RH + Cl_2 \xrightarrow[\text{light}]{\text{heat or}} RCl + HCl$$

Thermal or photochemical decomposition of a halogen molecule such as Cl_2 produces reactive chlorine atoms in the *initiation step*. These then collide with the alkane hydrocarbon, RH, to abstract a H· atom producing HCl and the organic free radical, R·. This radical then collides with a Cl_2 molecule to produce the alkyl halide product, RCl, along with another reactive chlorine atom, Cl·. These are the *propagation steps* which occur repeatedly, each sequence producing a product molecule, RCl, and a reactive species, Cl·, to continue the chain. This mechanism is called a *chain reaction* for this reason. These propagation steps continue until reactive radicals collide and combine to produce stable molecules. This is the *termination step,* which ends the chain process.

Initiation Step

$$Cl_2 \xrightarrow[\text{or } \Delta]{\text{light}} 2\ Cl\cdot$$

Propagation Steps

a) $Cl\cdot + RH \longrightarrow HCl + R\cdot$

b) $R\cdot + Cl_2 \longrightarrow RCl + Cl\cdot$

Termination Steps

$$2\ Cl\cdot \longrightarrow Cl_2$$

$$2\ R\cdot \longrightarrow R{-}R$$

$$Cl\cdot + R\cdot \longrightarrow R{-}Cl$$

Fig 21-2: Alkane halogenation by a free radical chain mechanism.

Note again that when several possible isomeric products can be formed, the product derived from the most stable (tertiary) free radical is favored:

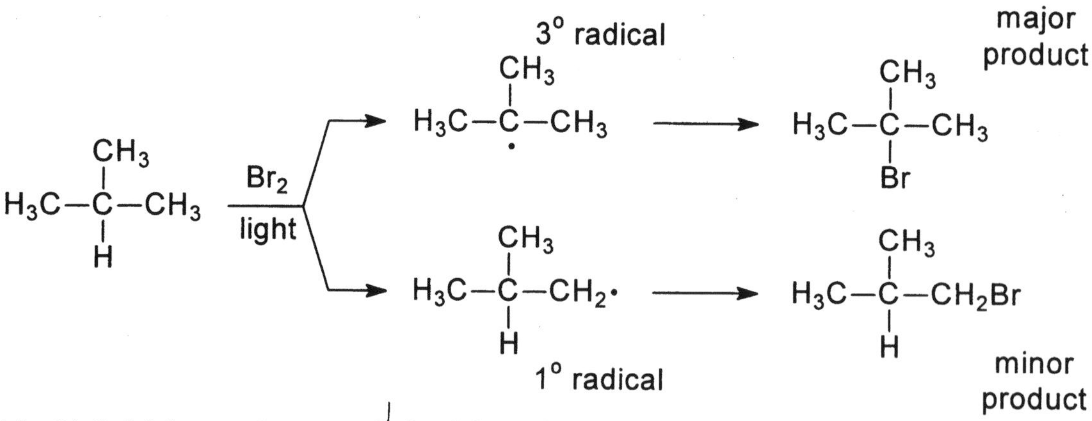

Fig 21-3: Major products are derived from the more stable free radical; minor products from the less stable.

Anti-Markovnikov Addition of HBr to Alkenes

The addition of HBr to simple alkenes normally proceeds by a carbocation mechanism to yield primarily the alkyl halide isomer derived from the more stable of the two possible carbocations. This is called the *Markovnikov* product after the Russian chemist who studied this reaction.

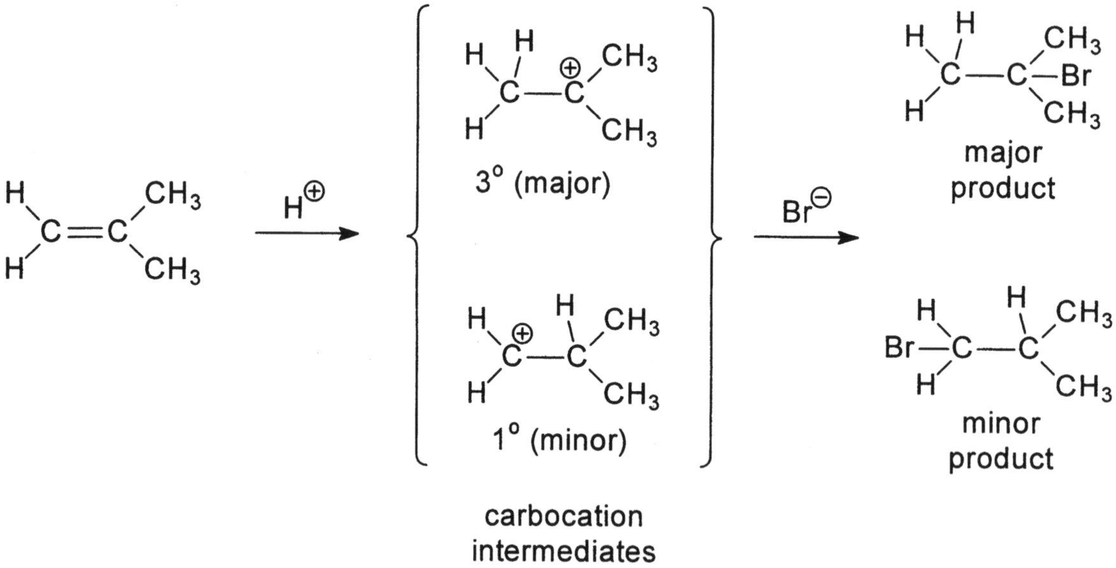

Fig 21-4: Markovnikov carbocation mechanism for the addition of HBr to isobutylene.

However, in the presence of a peroxide (which easily decomposes to produce free radicals) a free-radical chain mechanism preferentially occurs. This proceeds as shown below with the predominant alkyl bromide isomer being derived from the more stable of the two possible *free radical* intermediates. This produces the primary (1°) alkyl bromide as the major product in this example. Since this is the minor product obtained in the conventional carbocation (Markovnikov) mechanism, the product of the radical chain mechanism is referred to as the *anti-Markovnikov* product.

96

Initiation Steps

R—O—O—R $\longrightarrow$ 2 RO·

peroxide   peroxy radicals

RO· + HBr $\longrightarrow$ ROH + Br·

Propagation Steps

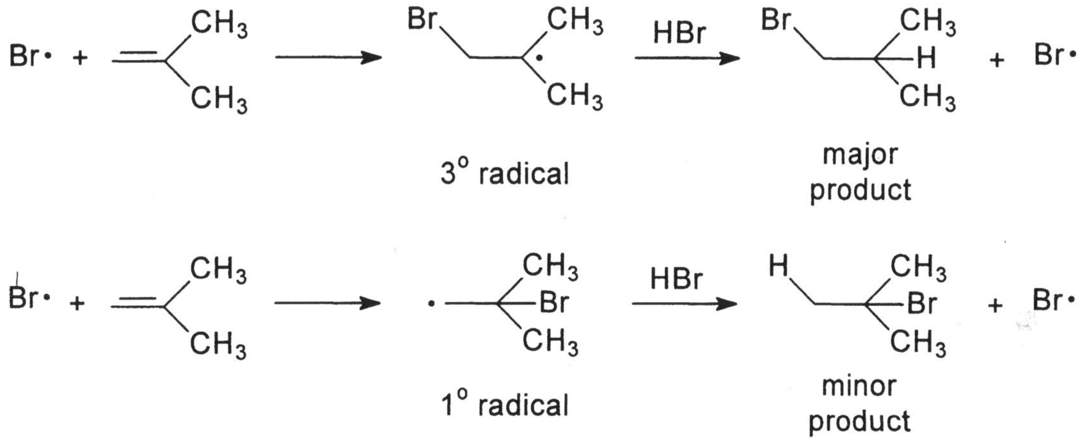

Termination Steps

2 Br· $\longrightarrow$ Br$_2$

Br· + R· $\longrightarrow$ RBr

Fig 21-5: Anti-Markovnikov free-radical chain mechanism for addition of HBr to isobutylene in the presence of peroxides.

A few things should be noted about the ionic and free-radical pathways for addition of HBr to alkenes:

1. A change in reaction mechanism occurred in the presence of peroxide. This mechanistic change has a direct consequence on the isomeric composition of the product mixture.
2. Only a trace of peroxide is needed, since the rate of the chain reaction is very fast compared to the competing, non-chain, carbocation process. (*Many* alkyl halide product molecules are produced for each peroxy radical initially produced.)
3. Both the ionic carbocation and the free-radical chain mechanisms give as major products the isomeric alkyl bromide derived from the *more stable* of two possible reactive intermediates derived from an unsymmetrical alkene. The difference in product mixture is due to the sequence of steps. The H$^+$ adds *first* in the carbocation mechanism to form the more stable carbocation; the Br· adds *first* in the radical mechanism to form the more stable radical.

Free-Radical Polymerization

Many polymers and plastics are produced by alkene polymerization initiated by a few free radicals. These are also chain reactions proceeding by the following mechanism:

1. Thermal or Photochemical Decomposition of an Initiator to Produce a Few Reactive Radicals:

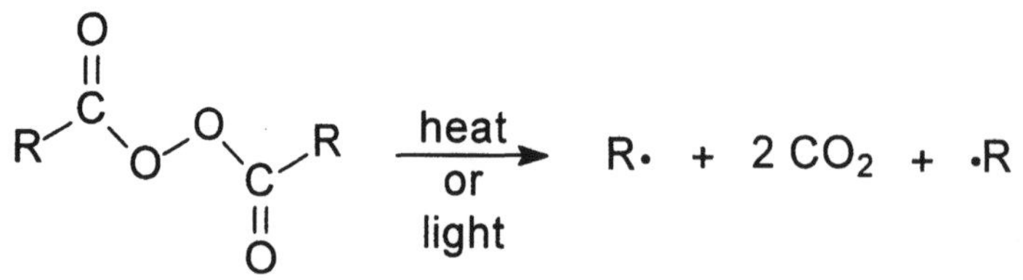

diacylperoxide

$$R-N=N-R \xrightarrow[\text{light}]{\text{heat or}} R\cdot + N_2 + \cdot R$$

azo compound

2. Propagation by Sequential Addition of Radicals to Alkenes Forming New Radicals of Longer Chain Length:

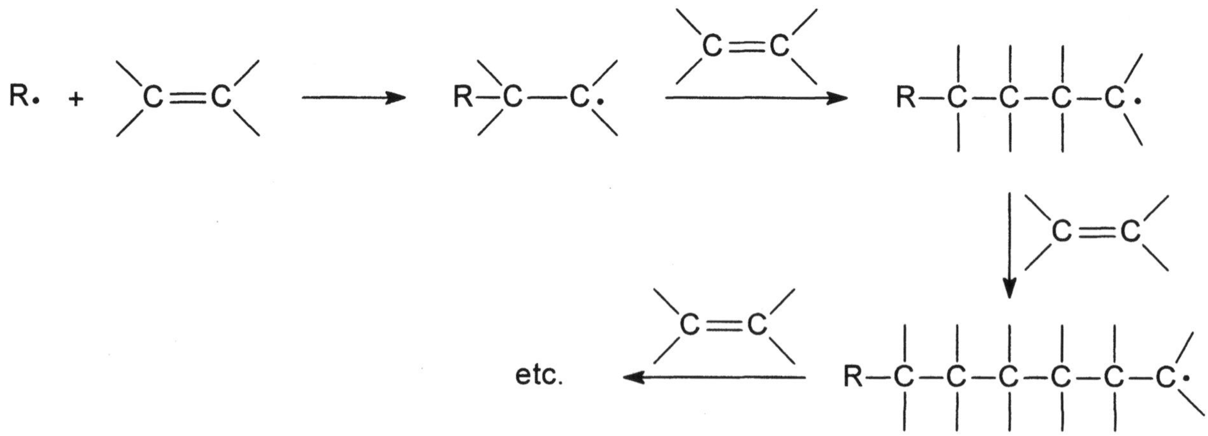

etc.

3. Chain Termination by Combination of any Pair of Free-Radicals:

$$R\cdot + \cdot R' \longrightarrow R-R'$$

98

Among the common plastics, polyethylene, polypropylene, PVC (polyvinylchloride), teflon (polytetrafluoroethylene), and many others are prepared by radical polymerization of the appropriately substituted alkene monomer. Note that these polymeric products are often called *polyolefins* since the monomers are olefins, an old name for alkenes.

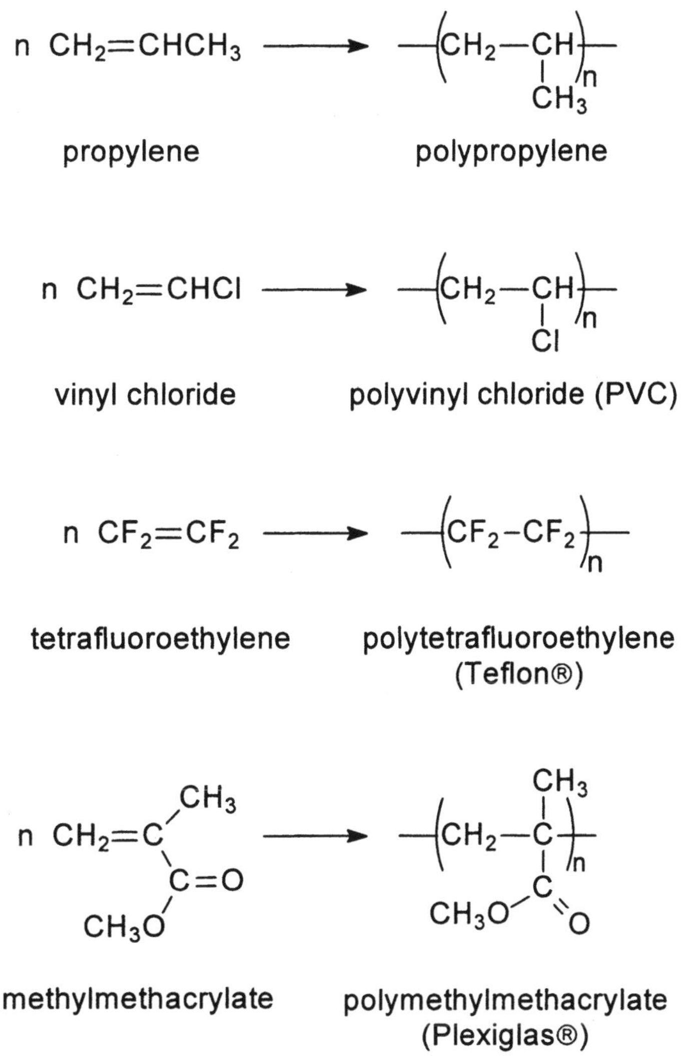

$$n \ CH_2{=}CHCH_3 \longrightarrow -\left(CH_2{-}\underset{\underset{CH_3}{|}}{CH}\right)_n-$$

propylene   polypropylene

$$n \ CH_2{=}CHCl \longrightarrow -\left(CH_2{-}\underset{\underset{Cl}{|}}{CH}\right)_n-$$

vinyl chloride   polyvinyl chloride (PVC)

$$n \ CF_2{=}CF_2 \longrightarrow -\left(CF_2{-}CF_2\right)_n-$$

tetrafluoroethylene   polytetrafluoroethylene (Teflon®)

methylmethacrylate   polymethylmethacrylate (Plexiglas®)

Fig 21-6: Some free-radical polymerization reactions.

Chapter 22
Alcohols and Ethers

Structure and Physical Properties

Alcohols have the general structure ROH. They are organic analogs of water, HOH, in which one of the hydrogen atoms is replaced by a carbon. Like water, alcohols form hydrogen bonds, which causes alcohol molecules to associate with each other:

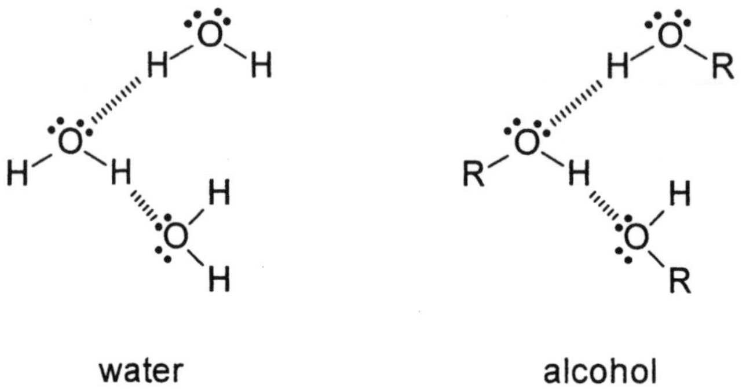

water   alcohol

Ethers have the general structure ROR′, in which the H atom of an alcohol is replaced by another carbon. Ethers are not capable of intermolecular hydrogen bonding to each other, since they have no oxygen-hydrogen bond. For this reason, alcohols have higher boiling points than ethers of similar molecular weight. This can be seen by comparing the boiling temperatures of the 2, 4, and 8 carbon alcohols and isomeric ethers: (Me, Et, and Bu refer to methyl, ethyl, and butyl.)

Alcohols	°C	Ethers	°C
EtOH	64.7°	Me-O-Me	-24.9°C
BuOH	117.7°	Et-O-Et	34.6°C
OctylOH	195°C	Bu-O-Bu	141°C

Fig 22-1: Boiling points of alcohols and ethers.

Water solubilities, however, tend to be similar for molecules containing either functional group, as indicated by the following data. This is because water (the solvent) can hydrogen bond to *either* alcohols or ethers (the solute).

Alcohols	gr/100 ml H_2O	Ethers	gr/100 ml H_2O
EtOH	∞	Me-O-Me	∞
BuOH	8.3	Et-O-Et	8
OctylOH	0.05	Bu-O-Bu	low

Fig 22-2: Water solubility of alcohols and ethers.

As in all homologous series of organic compounds, increasing the number of carbon atoms causes both boiling point and melting point to increase in regular fashion. And as the ratio of carbon atoms to oxygen atoms increases, the alcohol or ether becomes less polar, less water soluble, and more soluble in nonpolar solvents like hexane.

Alcohols

Synthesis of Alcohols

Alcohols can be prepared by hydration of alkenes in aqueous acid. The alkene accepts a proton (protonation) producing a carbocation which then reacts with water. The sequence is completed by proton loss (deprotonation.)

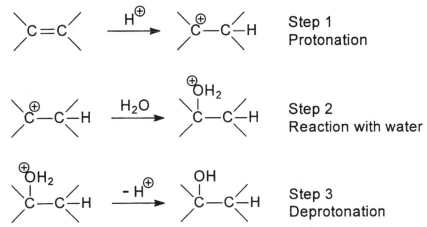

Fig 22-3: Mechanism of acid catalyzed hydration of an alkene to produce an alcohol.

Since this reaction proceeds through a carbocation intermediate, the predominant isomeric alcohol will derive from the more stable carbocation (Markovnikov isomer) when an *unsymmetrical* alkene is hydrated. Furthermore, the initially produced carbocation can be expected to rearrange by an alkide or hydride shift if a more stable carbocation can result, leading to an isomeric alcohol. A complex equilibrium of hydrations and dehydrations can occur in some cases to produce a mixture of isomeric alcohols and alkenes. Carbocations can also add to alkenes, further complicating the reaction mixture. All of these factors can lead to the production of undesired isomers, low yields of desired isomers, and complex product mixtures requiring elaborate separation and purification schemes.

101

To avoid these problems in the acid catalyzed hydration of olefins, several synthetic methods have been developed, intended to specifically produce high yields of either the Markovnikov or anti-Markovnikov alcohol isomers. These include methods (1) and (2) below, which are very important in practical synthetic chemistry.

Specific Methods of Alkene Hydration

1. Oxymercuration/Demercuration

   Reaction of the alkene with mercuric acetate, $Hg(OAc)_2$, followed by reduction with sodium borohydride ($NaBH_4$).

$$R-CH=CH_2 \xrightarrow[\text{2) NaBH}_4]{\text{1) Hg(OAc)}_2} \begin{array}{c} R-CH-CH_3 \\ | \\ OH \end{array}$$

Markovnikov
Isomer

2. Hydroboration

   Reaction with borane (BH_3) in Tetrahydrofuran (THF) followed by oxidation using basic hydrogen peroxide, H_2O_2.

$$R-CH=CH_2 \xrightarrow[\text{2) H}_2\text{O}_2, \text{ OH}^{\ominus}]{\text{1) THF:BH}_3} R-CH_2-CH_2OH$$

Anti-Markovnikov
Isomer

The Grignard Reaction

The *Grignard reaction* is one of the most important and useful synthetic reactions in organic chemistry. It is employed primarily to produce alcohols from ketones, aldehydes, esters, and epoxides. These are reacted with a *Grignard Reagent,* which is the magnesium salt of a carbanion, produced by reaction of an alkyl or aryl halide with magnesium metal. In the first step, an alkyl or aryl halide (usually a bromide or chloride) is reacted with magnesium turnings in dry ether. This is an electron transfer (redox) reaction in which two electrons lost by the magnesium are accepted by the organic halide. This instantly produces the *magnesium halide carbanion* mixed salt, or Grignard Reagent:

$$Mg + RX \longrightarrow \left(Mg^{+2} \; R-X^{-2} \right) \longrightarrow Mg^{+2} R^{-1} X^{-1}$$

Grignard Reagent

The purpose of this step is to produce the reactive carbanion, R⁻. This strong Lewis base is then reacted with the electron-deficient carbon atom (Lewis acid) of a suitable aldehyde, ketone, ester, or epoxide. As shown below, this gives rise to a particular 1°, 2°, or 3° alcohol after adding the reaction mixture to water. (Primary, 1°, secondary, 2°, and tertiary, 3°, refer to the number of carbon atoms bonded to the carbon carrying the OH group.) Notice that in reaction with aldehydes, epoxides, and ketones, the product derives directly from simple nucleophilic addition of the carbanion to the positively polarized carbonyl or ether carbon atom. Reaction with esters first involve substitution for the alkoxy group to produce a ketone, which then reacts further to produce the 3° alcohol shown. In the ester reaction, the final product alcohol contains *two* alkyl groups derived from the starting alkyl halide. All these reactions are spoken of as *reductive alkylations* since they all involve reduction to an alcohol along with introduction of one or more alkyl groups. (Note that the epoxide is in the same oxidation state as a *carbonyl* group.)

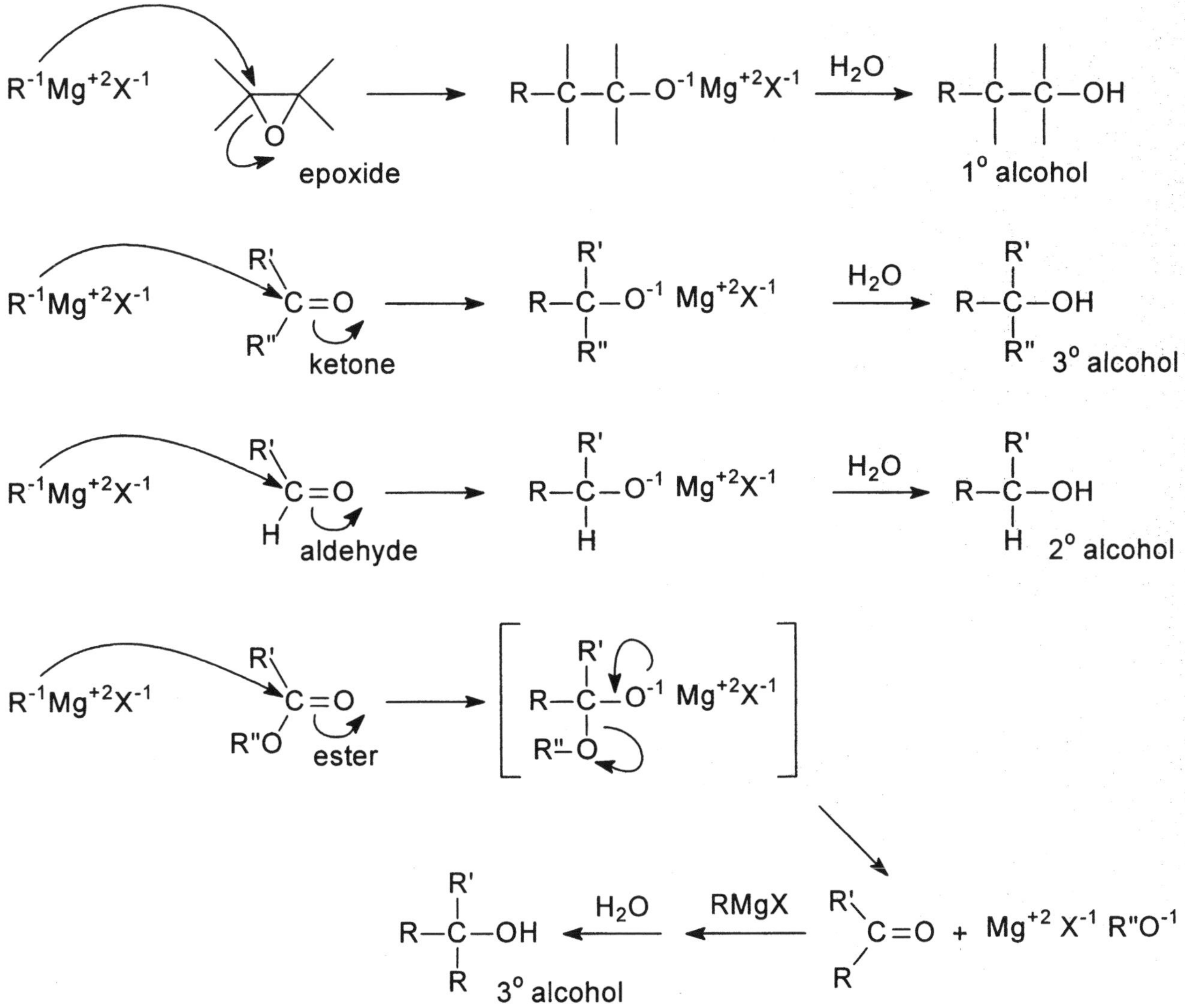

Fig 22-4: Grignard reactions: reductive alkylation to produce alcohols.

103

Here is a summary of important facts about Grignard reactions:

1. Alkyl or aryl halides first react with Mg metal to produce a Grignard reagent, which contains the reactive carbanion, R^-.

$$RX + Mg \longrightarrow R^{-1} \ Mg^{+2} X^{-1}$$

carbanion

2. The carbanion, R^-, then reacts as a strong nucleophile, attacking the electron-deficient carbon of aldehydes, ketones, esters, or epoxides. Upon hydrolysis the isolated product has undergone reductive alkylation to produce a 1°, 2°, or 3°, alcohol.

3. Since Grignard reagents are essentially salts of carbanions, they are also very strong bases and will react instantly with any available proton source in the environment. This acid-base reaction is very fast, and will neutralize and eliminate the carbanion. Such proton sources include all the functional groups in which a hydrogen atom is directly attached to a nitrogen, oxygen, or sulfur atom, or to an alkyne (due to the relatively high acidity of alkyne hydrogens). These groups include:

$$-OH, \quad -NH_2, \quad -NHR, \quad -COOH, \quad -SO_3H$$

$$-SH, \quad -CONH_2, \quad -C\equiv CH, \quad + \text{ others}$$

In general, these functional groups *cannot* be present in the reacting molecules if a Grignard reaction is to be carried out successfully. In practical terms, this means that only *unsubstituted* alkyl halides and an *unsubstituted* aldehyde, ketone, ester, or epoxide can be used for Grignard reactions.

4. Modified Grignard reagents are also employed in which 2 moles of lithium are used in place of 1 mole of magnesium. These produce so-called lithium Grignard reagents which react similarly to conventional magnesium Grignard reagents:

$$2\ Li\ +\ RX\ \longrightarrow\ R^{\ominus}Li^{\oplus}\ +\ Li^{\oplus}X^{\ominus}$$

Lithium
Grignard

Reactions of Alcohols

Nucleophilic Substitution

Nucleophiles are Lewis bases. These are molecules or ions that can donate an electron pair to form a covalent bond with an electron deficient center, which serves as a Lewis acid. Nucleophiles include anions like Cl^-, Br^-, I^-, hydroxide, OH^-, or alkoxides, RO^-, or neutral (uncharged) molecules like amines, NH_2R, water, HOH, or alcohols, ROH. Each of these has at least one pair of non-bonded electrons which can be donated to a Lewis acid to form a covalent bond. In general, the more reactive nucleophiles are the anions of the weakest acids, and the least reactive are the anions of the strongest acids. (Exceptions do occur, however; bromide and iodide ions are strong nucleophiles in substitution reactions with alkyl halides.) Conversely, the best (most reactive) leaving groups are the anions or conjugate bases of the strongest acids, and the poorest (least reactive) leaving groups are the conjugate bases of the weakest acids.

$$R{-}L\ +\ Nuc^{\ominus}\ \longrightarrow\ R{-}Nuc\ +\ L^{\ominus}$$

$$\left(\begin{array}{l} Nuc^{\ominus}\text{ is the nucleophile.} \\ L^{\ominus}\text{ is the leaving group} \end{array}\right)$$

Fig 22-5: Nucleophilic substitution reactions.

In alcohols, the leaving group is the OH^- ion, which is the conjugate base of H_2O, a relatively weak acid. For this reason, *the OH^- ion is a poor leaving group.* The rate of loss of this leaving group can be enhanced and reactivity increased, by *protonation*—carrying out the substitution in an acidic medium. When protonated, the leaving group becomes H_2O, the conjugate base of the hydronium ion, H_3O^+, which is a much stronger acid. *Thus, H_2O is a better (faster) leaving group than OH^-.*

$$ROH + Nuc^{\ominus} \longrightarrow R{-}Nuc + OH^{\ominus}$$

(SLOW: Leaving group is $OH^{\ominus}$)

$$ROH + H^{\oplus} \longrightarrow ROH_2^{\oplus}$$

$$ROH_2^{\oplus} + Nuc^{\ominus} \longrightarrow R{-}Nuc + H_2O$$

(FAST: Leaving group is H_2O)

Fig 22-6: Nucleophilic substitution for the OH group in neutral solution (top) and acidic solution (bottom).

Another very clever and commonly used approach to increase the rate of OH^- loss, is to employ a special reagent to chemically modify the OH group converting it to a *better leaving group*. Here, the OH group is converted to the ester of a strong acid, a sulfonic acid, using the reagent p-toluene sulfonyl chloride or *tosyl chloride*.

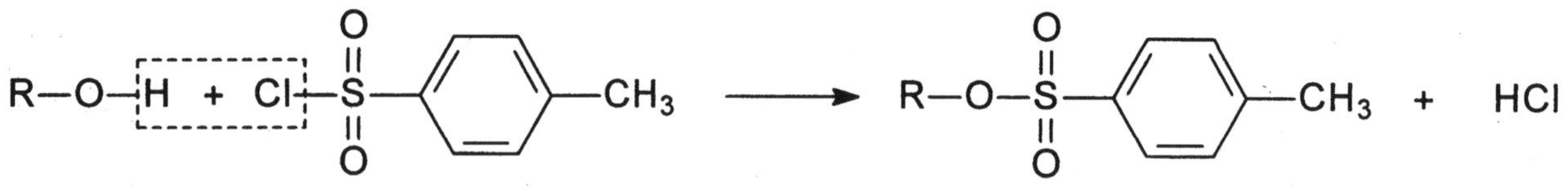

R—O⫶H + Cl⫶—S(=O)(=O)—⟨C₆H₄⟩—CH₃ ⟶ R—O—S(=O)(=O)—⟨C₆H₄⟩—CH₃ + HCl

alcohol   tosyl chloride   tosyl ester

Now the tosylate anion is the leaving group. Since this is the anion of a very strong acid (p-toluene sulfonic acid), substitution by a nucleophile, Nuc^-, proceeds very quickly and easily to produce a high yield of the desired product. In this way a *poor* (unreactive) leaving group is converted into a *good* (reactive) leaving group to enhance reactivity in nucleophilic substitution reactions. (The reagent methyl sulfonyl chloride, also called mesyl chloride, MsCl, can be used in much the same way as tosyl chloride.)

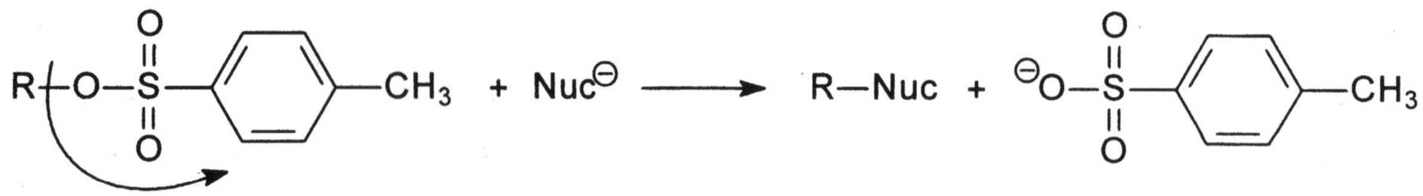

R—O—S(=O)(=O)—⟨C₆H₄⟩—CH₃ + $Nuc^{\ominus}$ ⟶ R—Nuc + $^{\ominus}$O—S(=O)(=O)—⟨C₆H₄⟩—CH₃

tosyl ester   nucleophile   substitution product   tosylate anion

Ethers

Synthesis of Ethers

The major reaction used for the preparation of ethers is the *Williamson ether synthesis* in which the oxyanion of an alcohol, RO⁻ is reacted by an S_N2 mechanism with an alkyl halide or another alkyl derivative containing a good leaving group:

$$RO^{\ominus} Na^{\oplus} + R'{-}X \xrightarrow{S_N2} R{-}O{-}R' + Na^{\oplus} X^{\ominus}$$

$$\text{ether}$$

Reactions of Ethers

Generally, ethers are quite unreactive and participate in very few chemical reactions. They can, however, be cleaved to alcohols and alkyl halide fragments by heating at elevated temperature under pressure with concentrated HBr or HI. (The initially formed alcohol reacts further under these conditions to produce a second molecule of alkyl halide, as shown.)

$$R{-}O{-}R' \xrightarrow{HX} R{-}O{-}H + R'{-}X$$

$$\xrightarrow[HX]{} R{-}X + H_2O$$

$$\text{alkyl halides}$$

An exception to the low general reactivity of ethers is the three-membered ring cyclic ethers, called *oxiranes* or *epoxides*. These can be produced by reaction of an alkene with per-acids such as *magnesium monoperoxyphthalate* (MMPP), or by reaction of halohydrins with base:

alkene   MMPP   epoxide

The MMPP reaction involves the insertion of a peroxy oxygen into the double bond. Reaction of the halohydrin proceeds via intramolecular S_N2 substitution.

Due to angle and torsional strain, these epoxides undergo easy ring opening, resulting from S_N2 attack on one of the ring carbon atoms.

Fig 22-7: Epoxide ring opening reactions.

(The ring can also be opened by a carbocation mechanism in an acid medium if a relatively stable 3° carbocation is produced.)

Chapter 23
Addition and Substitution Reactions of Aldehydes and Ketones

The carbon-oxygen double bond, $\diagdown C = O$, called the *carbonyl group* is one of the most important and widely distributed atomic groupings in organic chemistry. It occurs in a range of functional groups including *ketones* and *aldehydes,* as well as *carboxylic acids* and their derivatives such as *esters* and *amides*. These constitute a significant fraction of all organic functional groups and occur in myriad compounds of both biological and industrial importance.

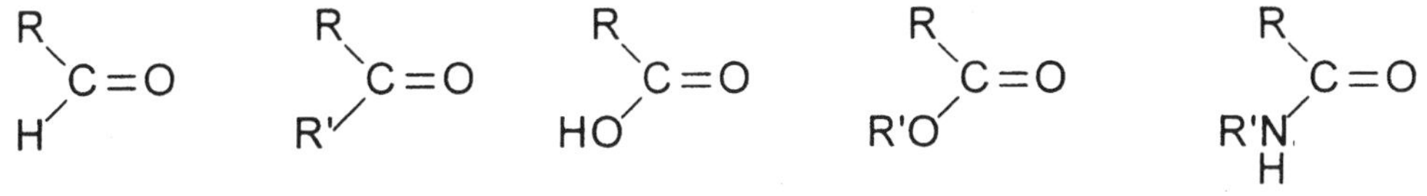

| aldehyde | ketone | carboxylic acid | ester | amide |

The most important fact about the carbonyl group and its reactions is that it possesses a permanent dipole as shown:

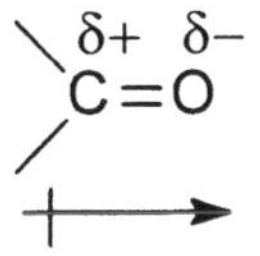

The electron cloud is pulled more strongly toward the electronegative oxygen atom than toward the carbon. This results in charge delocalization as shown, with the oxygen bearing a partial negative charge and the carbon end having a partial positive charge.

This electron deficiency and positive character at the carbonyl carbon atom causes it to behave as a Lewis acid (electron pair acceptor), and, is thus readily attacked by nucleophiles (Nuc⁻), which are Lewis bases (electron pair donors). *The great majority of all carbonyl reactions are initiated by this Lewis acid-base reaction.* The oxy-anion produced then acquires a proton from the reaction solvent or during product isolation (work-up). The product of this reaction may subsequently lose water when the nucleophile is an amine or amine derivative. In such cases the overall reaction is an *addition* followed by an *elimination,* which results in a *substitution.*

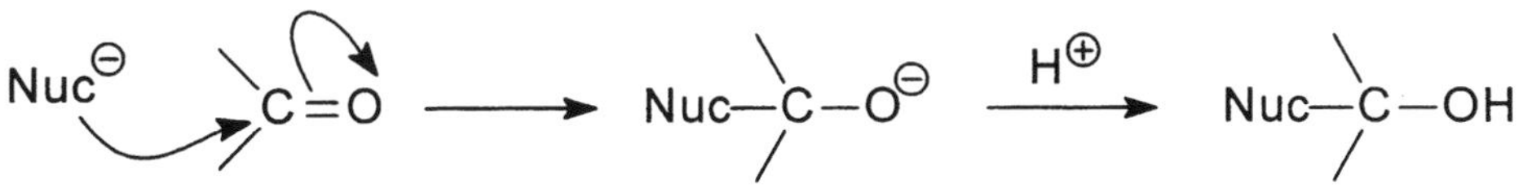

oxy-anion

Listed here are a number of important reactions of ketones and aldehydes which proceed by the mechanism described above:

1. Hydration (Reaction with Water):

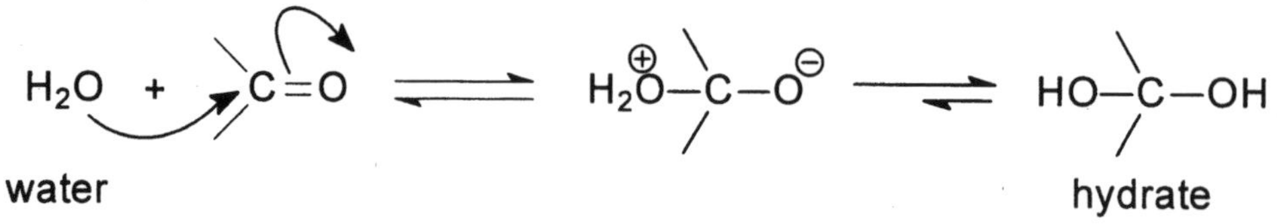

$$H_2O \; + \; \diagup\!\!\!\!C{=}O \; \rightleftharpoons \; H_2\overset{\oplus}{O}{-}\!\underset{\diagup}{C}{-}O^{\ominus} \; \rightleftharpoons \; HO{-}\!\underset{\diagup}{C}{-}OH$$

water   hydrate

2. Hemiacetal Formation (Reaction with Alcohols):

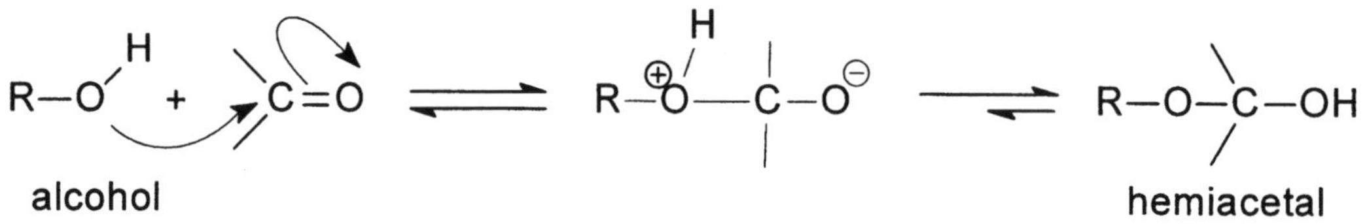

$$R{-}O\overset{H}{\diagup} \; + \; \diagup\!\!\!\!C{=}O \; \rightleftharpoons \; R{-}\overset{\oplus}{O}\overset{H}{-}\!\underset{\diagup}{C}{-}O^{\ominus} \; \rightleftharpoons \; R{-}O{-}\!\underset{\diagup}{C}{-}OH$$

alcohol   hemiacetal

3. Schiff Base Formation (Reaction with 1° Amine Followed by Dehydration):

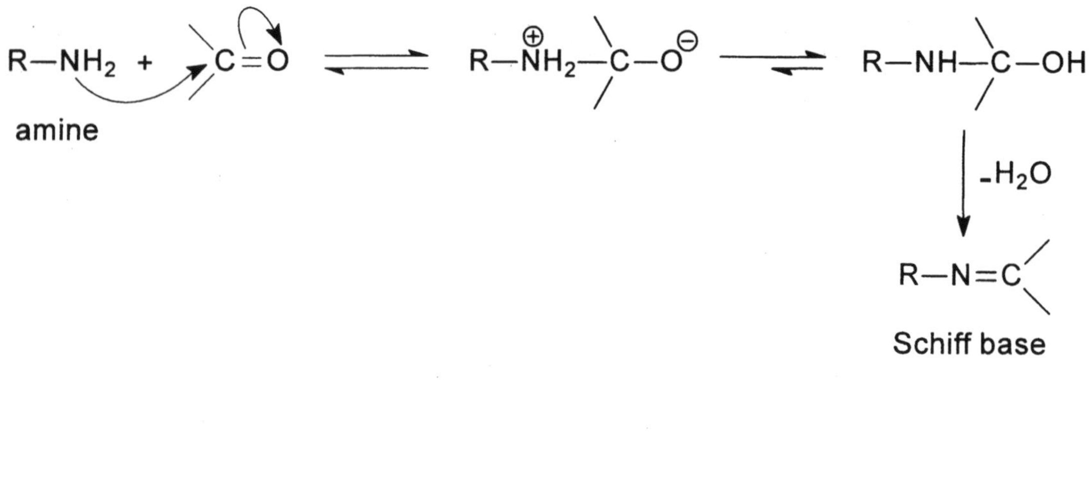

$$R{-}NH_2 \; + \; \diagup\!\!\!\!C{=}O \; \rightleftharpoons \; R{-}\overset{\oplus}{N}H_2{-}\!\underset{\diagup}{C}{-}O^{\ominus} \; \rightleftharpoons \; R{-}NH{-}\!\underset{\diagup}{C}{-}OH$$

amine

$$\downarrow {-}H_2O$$

$$R{-}N{=}C\diagdown^{\diagup}$$

Schiff base

110

4. Formation of Hydrazones and Oximes (Reaction with Hydrazines or Hydroxylamine, Followed by Dehydration):

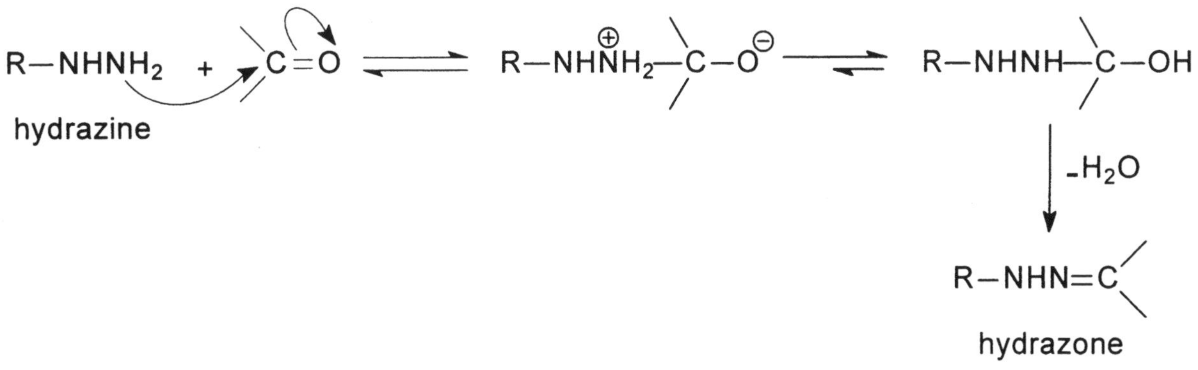

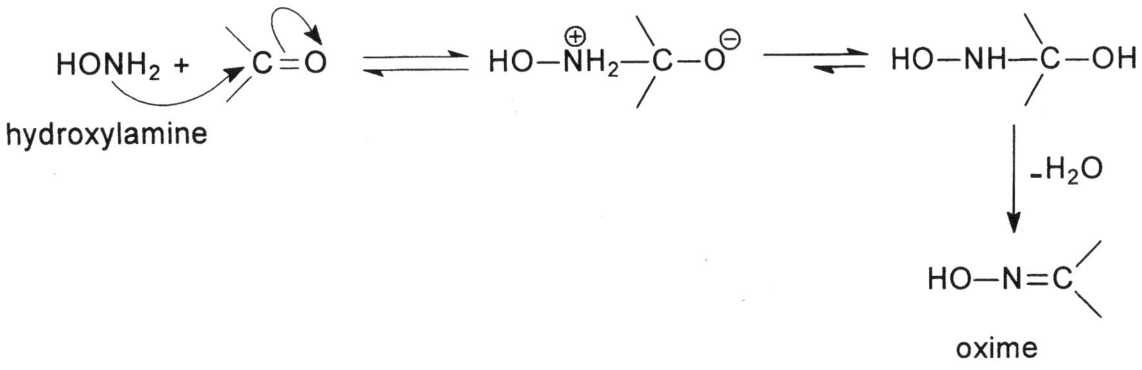

The initial step in each of these reactions is attack of a nucleophile on the carbonyl carbon. In reactions 3 and 4, above, this is followed by dehydration (loss of water), resulting in a substitution.

Prior to the advent of modern instrumental methods, oximes and hydrazones were routinely used for functional group analysis to establish the presence of an aldehyde or ketone in an organic compound. Ability to form a hydrazone and oxime derivative was strong evidence for the presence of the aldehyde or ketone functional group in a newly prepared or isolated compound. A unique structure assignment could often be made by comparing the melting points of the isolated derivatives with tabulated handbook values.

Chapter 24
Oxidation and Reduction Reactions of Carbonyl Compounds

Oxidation and reduction processes are defined by any one or more of the three transfer events shown:

OXIDATION	REDUCTION
Gain of oxygen	Loss of oxygen
Loss of hydrogen	Gain of hydrogen
Loss of electrons	Gain of electrons

      Oxidation and reduction are complimentary processes and always occur simultaneously: Gain of oxygen atoms by one molecule necessarily results in a loss of oxygen atoms by another. The same holds true for loss and gain of hydrogen atoms or of electrons. This is because chemical reactions must always comply with the basic principles of non-transmutability of elements and conservation of mass and charge.

      Functional groups can be converted to others in higher or lower oxidation states by oxidation/reduction, or *redox,* reactions. Successive oxidation/reduction reactions produce a redox series. For the oxygen-containing functional groups, which include alcohols, ketones and aldehydes, and carboxylic acids, two such series are:

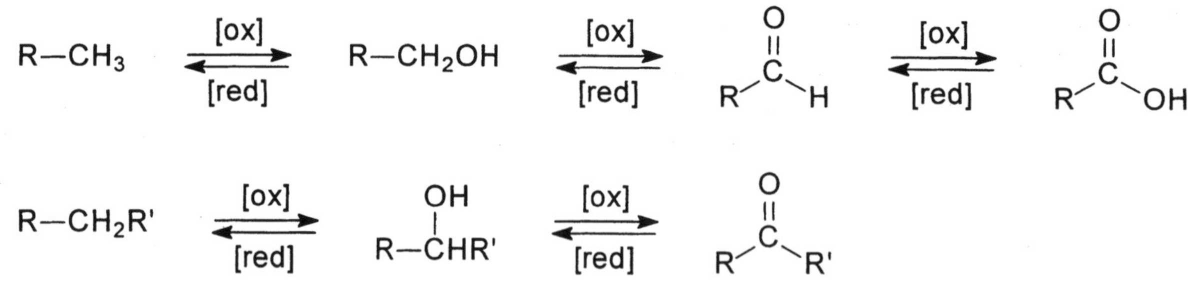

Reduction Mechanisms (Ketones → 2° Alcohols)

Consider the conversion of a ketone to a 2° alcohol:

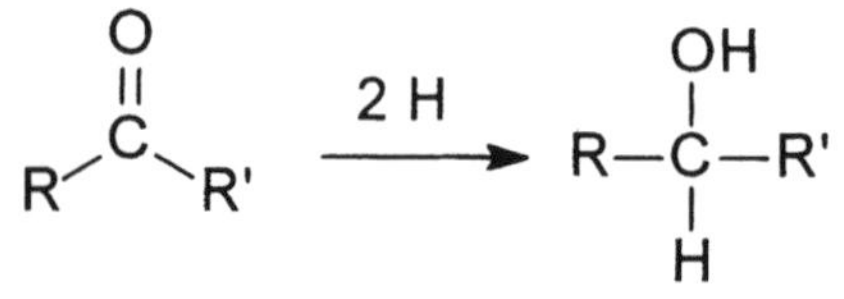

This is clearly a reduction since the ketone has accepted (gained) 2 hydrogen atoms. This process can be effected by three different mechanisms:

1. Hydride (H⁻) Attack Followed by Proton (H⁺) Transfer
    The regents sodium borohydride, $NaBH_4$, or lithium aluminum hydride, $LiAlH_4$, are commonly used to reduce ketones by this mechanism. These reagents, called *hydride donors,* transfer a hydride ion (H⁻) to the ketone followed by proton transfer from the solvent or in a later step.

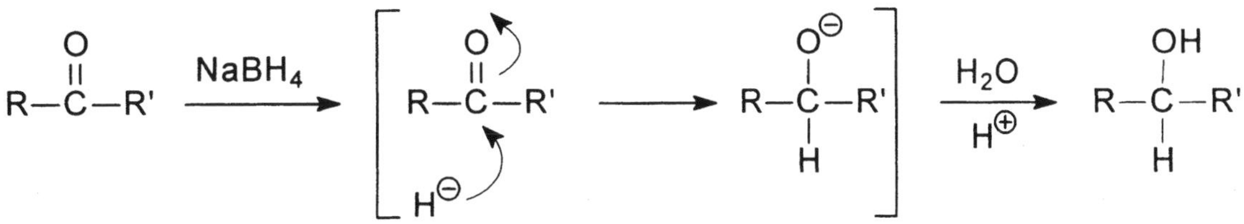

hydride attacks at
carbonyl carbon

2. Catalytic Hydrogenation
    Two hydrogen atoms can be transferred to the ketone carbonyl group by use of H_2 gas with a heavy metal catalyst. This reaction is analogous to the catalytic hydrogenation of alkenes to produce alkanes. The catalyst helps to break the carbonyl pi bond and the H_2 molecule into two hydrogen atoms so the addition reaction can proceed more easily:

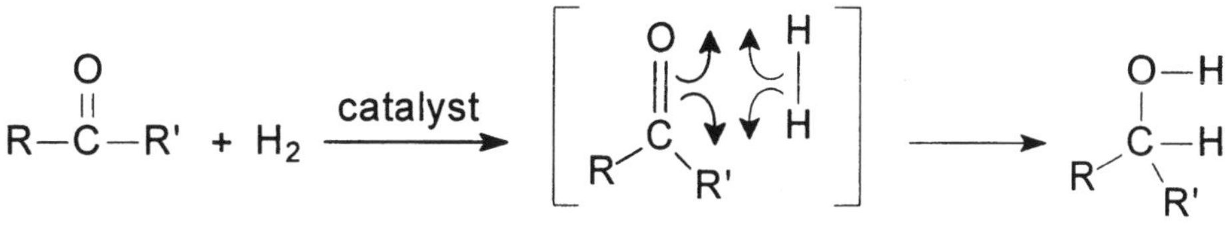

hydrogen atom transfer
at catalyst surface

3. Dissolving Metal Reduction

Electron transfer from an alkali metal, like sodium, followed by protonation can be repeated twice to effect the reduction of a ketone to an alcohol:

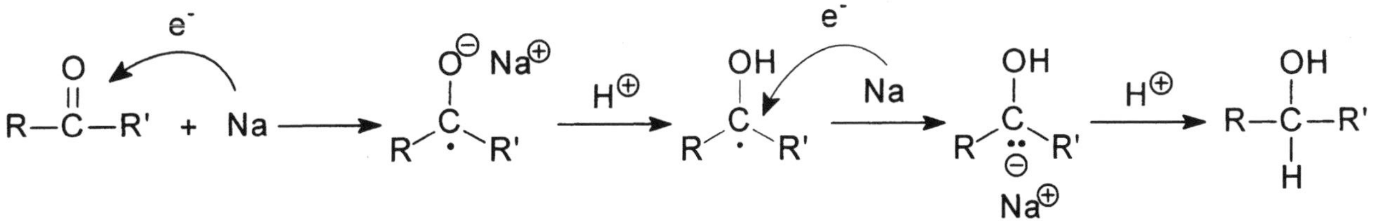

113

Thus, the reduction of a ketone to a 2° alcohol can be effected by three different mechanisms using a variety of reagents. In each case the net result is the addition of two hydrogen atoms to the ketone. (Similar mechanisms can transform aldehydes to the corresponding 1° alcohols):

REDUCTION MECHANISMS

Hydride Transfer	$(H^- + H^+)$ equals 2 H atoms
Catalytic Hydrogenation	$(H\cdot + H\cdot)$ equals 2 H atoms
Dissolving Metal	$(2e^- + 2H^+)$ equals 2 H atoms

Some examples are shown below to indicate the use and specificity of these various reduction methods. Note that sodium borohydride is able to reduce only aldehydes and ketones, while the more active reagent, lithium aluminum hydride, will also reduce esters and other carbonyl-containing functional groups. Note as well that hydride donors will reduce the polar carbonyl group, but not non-polar alkenes, which typically require catalytic hydrogenation for reduction to alkanes.

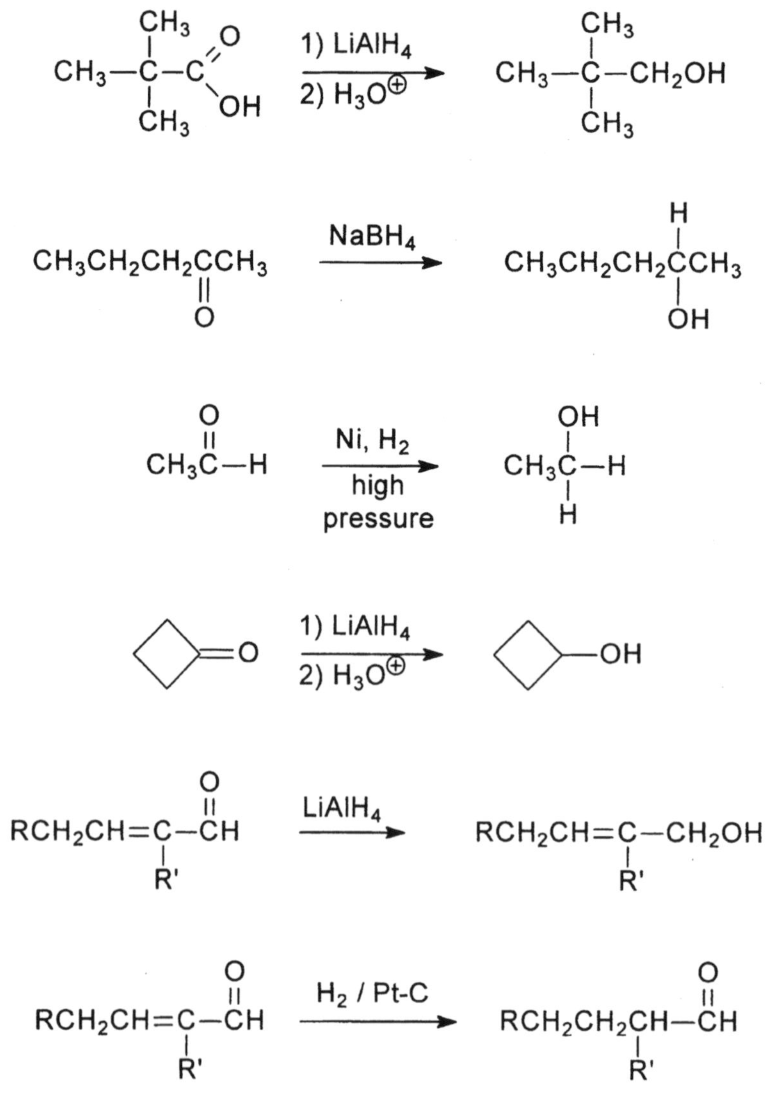

114

Oxidation of Alcohols to Aldehydes and Ketones

    Common oxidizing agents used for converting alcohols to other functional groups in higher oxidation states include potassium permanganate, $KMnO_4$, and chromium trioxide, CrO_3, and the chromate CrO_4^{-2}, and dichromate, $Cr_2O_7^{-2}$, ions. These contain manganese in the $+7$ and chromium in the $+6$ oxidation states. (These oxidation states are calculated by assuming oxygen to have a charge of -2. Manganese and chromium must then be assigned charges of $+7$ and $+6$ respectively, to account for the overall charge on the molecule or complex ion.) Oxidation of 1° alcohols generally produces carboxylic acids since the initially produced aldehyde is especially sensitive to continued oxidation. A particular oxidizing reagent, however, called *pyridinium chlorochromate*, or PCC, can specifically convert 1° alcohols to aldehydes in good yield. Some examples of these oxidations follow:

$$CH_3CH_2CH_2CH_2OH \xrightarrow[H_2SO_4]{K_2Cr_2O_7} CH_3CH_2CH_2C\overset{\displaystyle O}{\underset{\displaystyle OH}{}}$$

$$(CH_3CH_2)_2\overset{\displaystyle CH_3}{\underset{\displaystyle |}{C}}CH_2OH \xrightarrow[CH_2Cl_2]{PCC^*} (CH_3CH_2)_2\overset{\displaystyle CH_3}{\underset{\displaystyle |}{C}}-C\overset{\displaystyle O}{\underset{\displaystyle H}{}}$$

* pyridinium chlorochromate

115

Chapter 25
Carboxylic Acids and Derivatives

Carboxylic acids have the general structural formula shown below. This functional group derives its name from *carb*onyl and hydr*oxyl:*

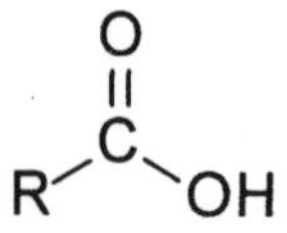

$$\underset{R}{\overset{\displaystyle O}{\underset{}{\overset{\|}{C}}}}\!\!-OH$$

Carboxylic acids are named with the suffix *-ic* or *-oic,* followed by the word *acid.* Here is a sampling of some members of this class of compounds, described using both systematic and common names:

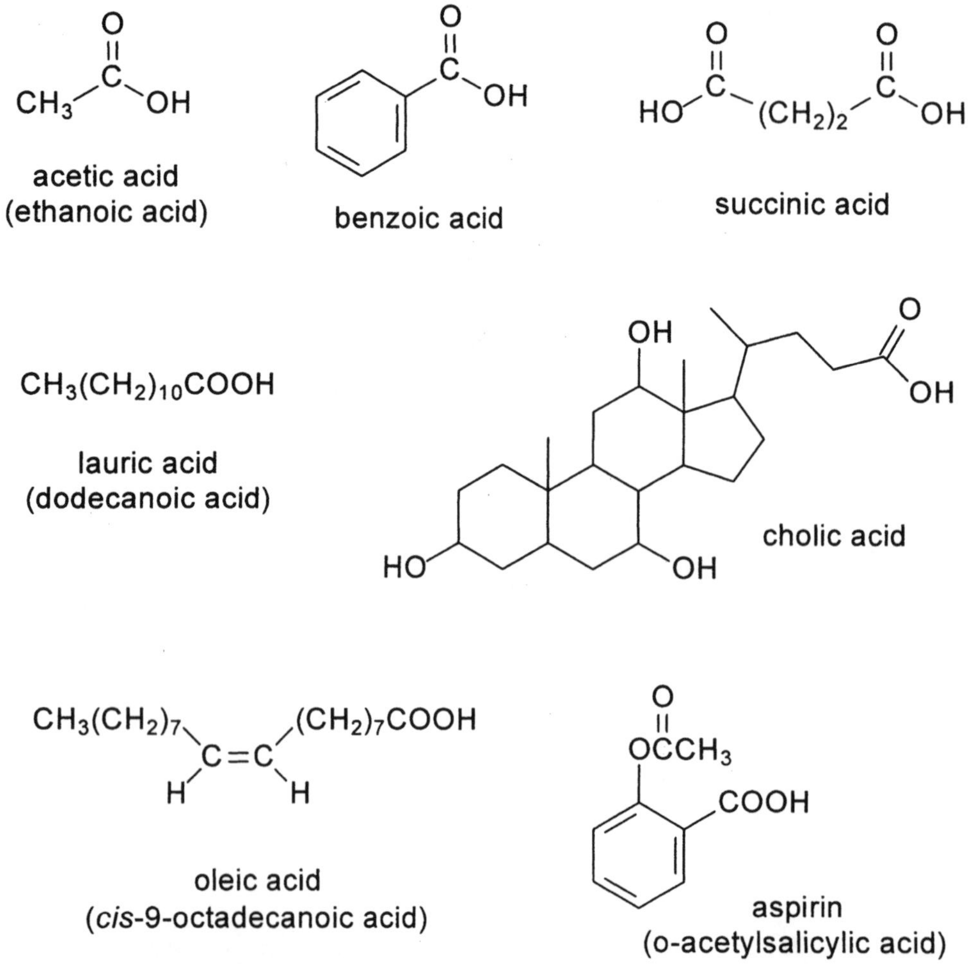

acetic acid
(ethanoic acid)

benzoic acid

succinic acid

$CH_3(CH_2)_{10}COOH$

lauric acid
(dodecanoic acid)

cholic acid

oleic acid
(*cis*-9-octadecanoic acid)

aspirin
(o-acetylsalicylic acid)

Fig 25-1: Some representative carboxylic acids.

A few carboxylic acids have been well known since the early days of organic chemistry. They are important in biochemical processes and figure prominently in ancient crafts and industries, such as wine and vinegar making, and soap production.

The primary characteristic property of carboxylic acids is *acidity*. These are the acids of organic chemistry. Ionization yields a proton and the *carboxylate anion*. (The suffix *-ate* is used to describe the anion.)

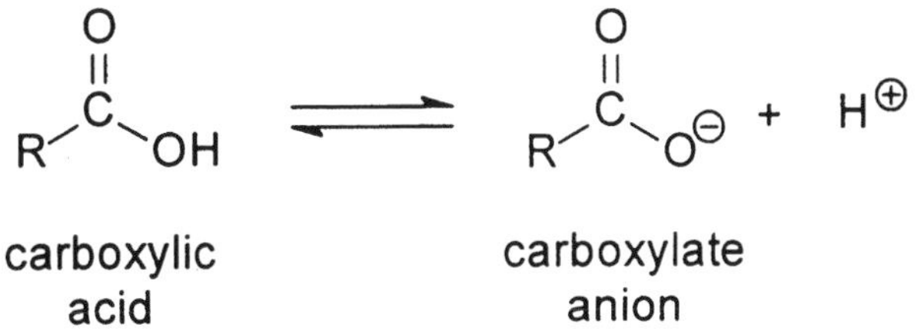

carboxylic
acid

carboxylate
anion

Carboxylic acids are relatively weak acids compared to the inorganic mineral acids such as H_2SO_4 or HCl. In dilute water solution they generally ionize to the extent of only a few percent. Most carboxylic acids will produce a pH between 2 and 4 when dissolved in water at typical concentrations.

Production of Carboxylic Acids

Exhaustive oxidation of 1° alcohols produces a carboxylic acid as the end product:

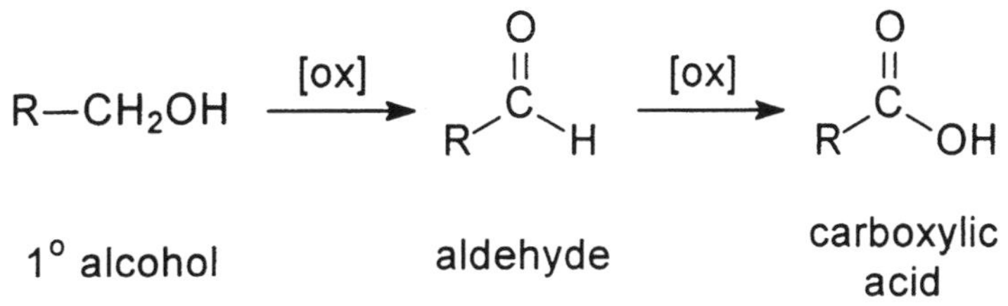

1° alcohol   aldehyde   carboxylic
acid

They can also be obtained by hot permanganate oxidative cleavage of alkenes possessing one hydrogen atom directly linked to the double bond, or by oxidative cleavage of alkynes:

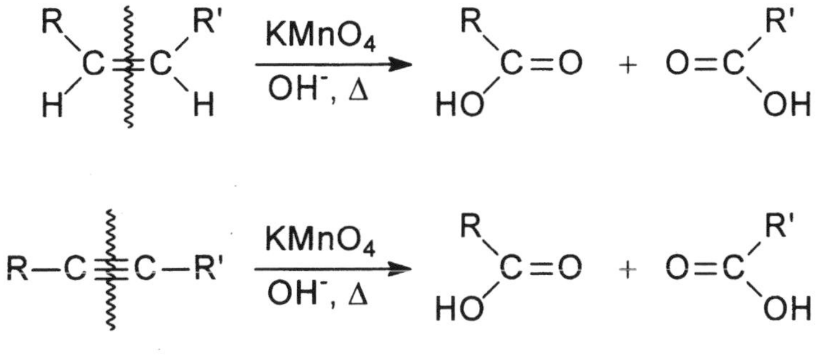

Grignard reagents can be reacted with CO_2 (dry ice) to produce carboxylic acids from alkyl halides. This reaction is mechanistically analogous to the reaction of Grignard reagents with the electron-deficient

117

carbonyl carbon of aldehydes and ketones. (See Chapter 22). Here the carbanion attacks the positively polarized carbon of carbon dioxide, which directly produces the carboxylate anion.

$$R\text{—}X \xrightarrow{\text{Mg}} RMgX \xrightarrow{\text{CO}_2} \underset{\text{Mg}^{+2}X^{-1}}{R\text{—}\overset{\overset{\displaystyle O}{\|}}{C}\text{—}O^{-1}} \xrightarrow{\text{H}_3\text{O}^+} R\text{—}\overset{\overset{\displaystyle O}{\|}}{C}\text{—}OH$$

$$\left(R^{\ominus} \underset{\delta-}{\overset{\delta-}{\overset{\displaystyle O}{\underset{\displaystyle O}{\|\|C\|\|}}}}{}_{\delta+} \longrightarrow R\text{—}\overset{\overset{\displaystyle O}{\|}}{C}\text{—}O^{\ominus} \right)$$

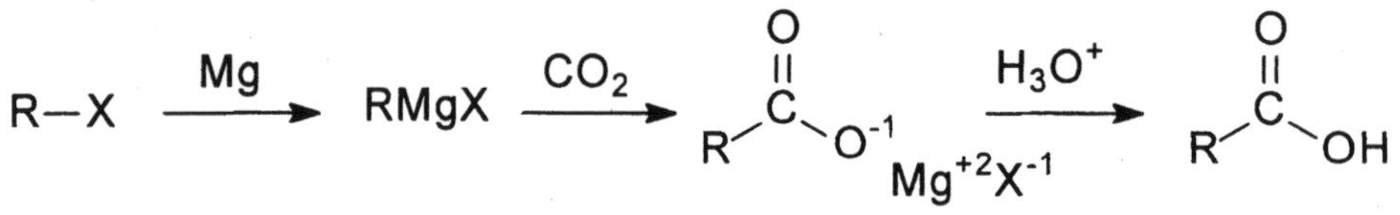

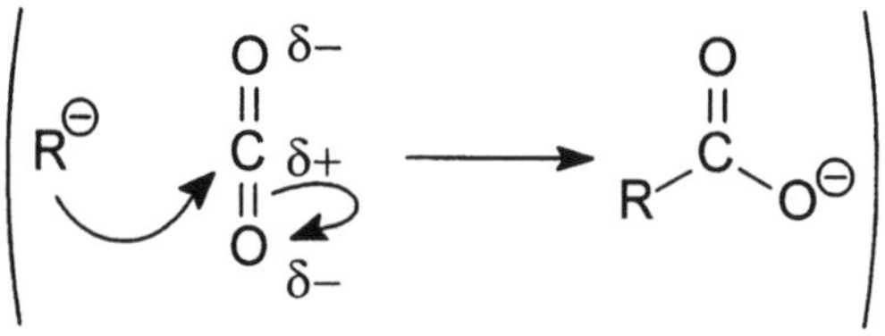

Reactions of Carboxylic Acids

Acid-Base Reactions:

Like any acid, carboxylic acids react with bases such as sodium or potassium hydroxide or amines to form salts. Since the carboxylate salt is an ionic product, the water solubility of carboxylic acids generally increases dramatically upon reaction with bases. In fact, a positive qualitative test for the presence of a carboxylic acid is the observation of a water-insoluble compound becoming freely soluble upon addition of a suitable base, usually sodium bicarbonate.

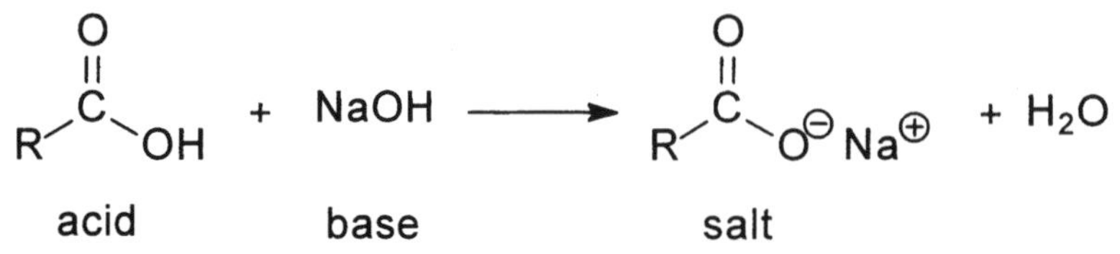

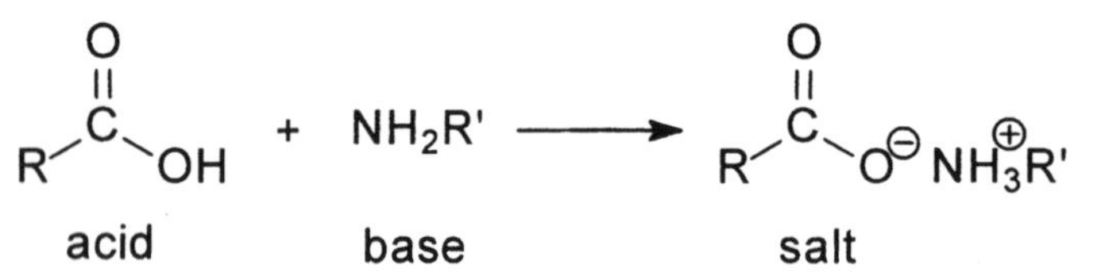

$$\underset{\text{acid}}{R\text{—}\overset{\overset{\displaystyle O}{\|}}{C}\text{—}OH} + \underset{\text{base}}{\text{NaOH}} \longrightarrow \underset{\text{salt}}{R\text{—}\overset{\overset{\displaystyle O}{\|}}{C}\text{—}O^{\ominus}\text{Na}^{\oplus}} + \text{H}_2\text{O}$$

$$\underset{\text{acid}}{R\text{—}\overset{\overset{\displaystyle O}{\|}}{C}\text{—}OH} + \underset{\text{base}}{\text{NH}_2\text{R}'} \longrightarrow \underset{\text{salt}}{R\text{—}\overset{\overset{\displaystyle O}{\|}}{C}\text{—}O^{\ominus}\text{NH}_3^{\oplus}\text{R}'}$$

This property of solubility in aqueous base is often utilized to separate carboxylic acids from non-acidic (neutral) organic compounds during purification of crude reaction mixtures. The carboxylate anion is formed by reaction of the carboxylic acid with hydroxide ions in water. The ionic carboxylate salt is then extracted into water, while the nonionic neutral organic impurities remain dissolved in a water-immiscible organic solvent. This procedure is commonly conducted using a *separatory funnel*. After separation the pure carboxylic acid can then be regenerated by reaction with a strong acid like HCl and isolated by filtration or re-extraction into an organic solvent.

118

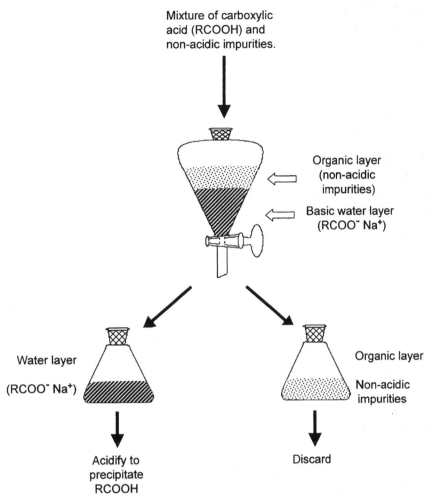

Fig 25-2: Purification of a carboxylic acid based on water solubility of the carboxylate salt.

Carboxylic Acid Derivatives

Substitution for the OH group of a carboxylic acid by various other heteroatomic groups produces the family of carboxylic acid derivatives:

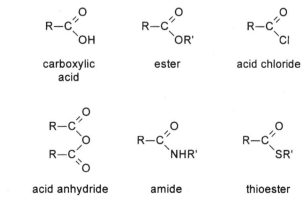

Fig 25-3: Carboxylic acid and carboxylic acid derivatives.

These comprise one of the most important families of functional groups in organic chemistry. They can be inter-converted by acyl (carbonyl) substitution reactions, in general described by the following equation:

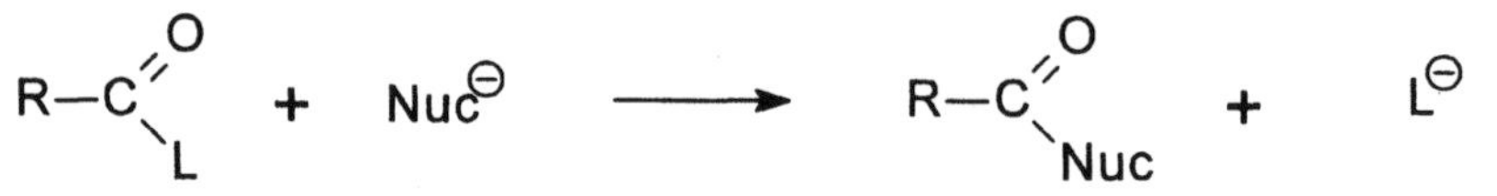

$$R-C\overset{O}{\underset{L}{\big<}} \quad + \quad Nuc^{\ominus} \quad \longrightarrow \quad R-C\overset{O}{\underset{Nuc}{\big<}} \quad + \quad L^{\ominus}$$

Here L is the leaving group and Nuc$^-$ is the entering nucleophile. Examples of these reactions include ester and amide hydrolysis, which are very important in both organic and biochemistry. In these reactions the leaving groups are an alcohol and an amine respectively, and the attacking nucleophile is water:

Ester Hydrolysis: $\quad R-C\overset{O}{\underset{OR'}{\big<}} \quad + \quad H_2O \quad \longrightarrow \quad R-C\overset{O}{\underset{OH}{\big<}} \quad + \quad R'OH$

Example:

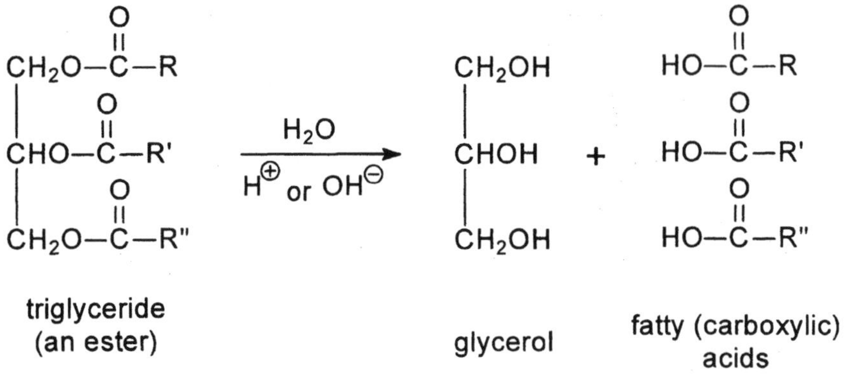

triglyceride
(an ester)   glycerol   fatty (carboxylic) acids

Amide Hydrolysis: $\quad R-C\overset{O}{\underset{NHR'}{\big<}} \quad + \quad H_2O \quad \longrightarrow \quad R-C\overset{O}{\underset{OH}{\big<}} \quad + \quad NH_2R'$

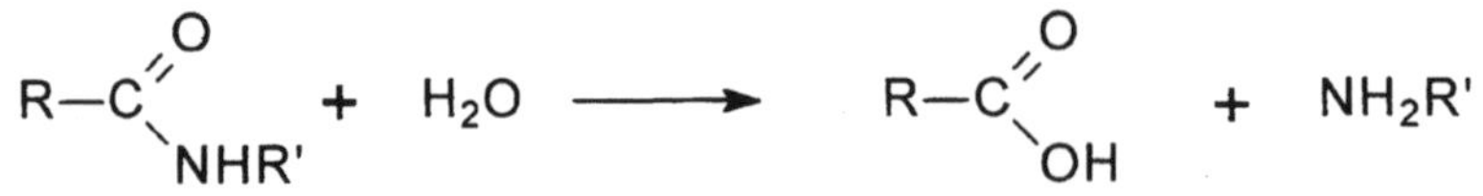

Example:

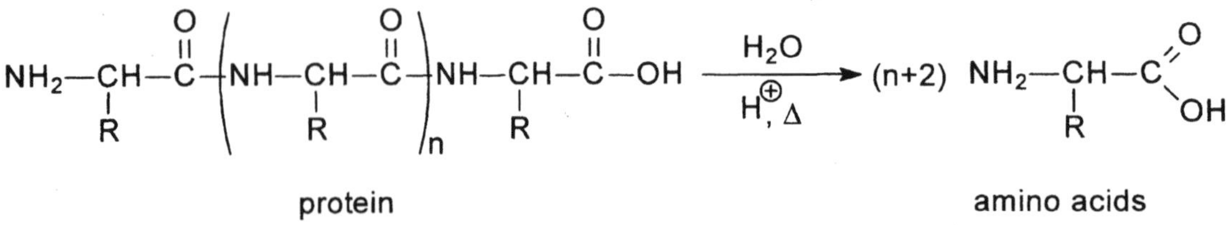

protein   amino acids

Fig 25-4: Examples of important substitution reactions of carboxylic acid derivatives. Top: Ester hydrolysis. Bottom: Amide hydrolysis.

These carbonyl substitutions are the most important reactions of carboxylic acids and their derivatives. Here are a few more examples of this type of reaction:

Deactivation of Penicillin

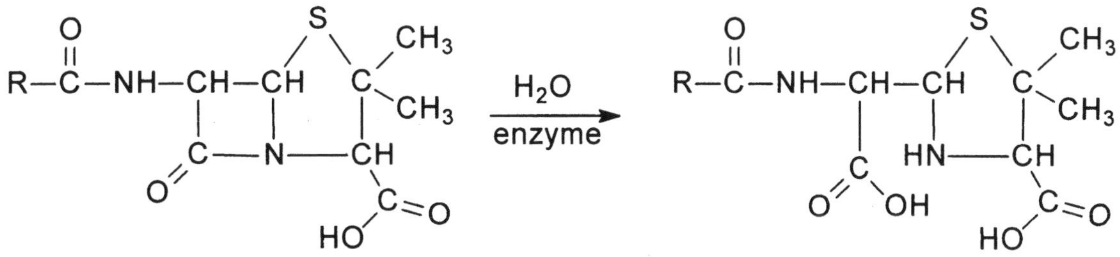

penicillin → penicilloic acid

Gabriel Synthesis of Amines

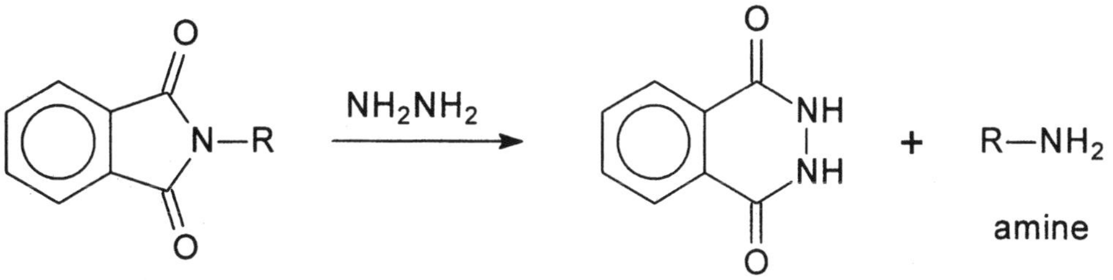

N-alkylphthalimide → + R—NH₂ (amine)

Synthesis of Barbituric Acid

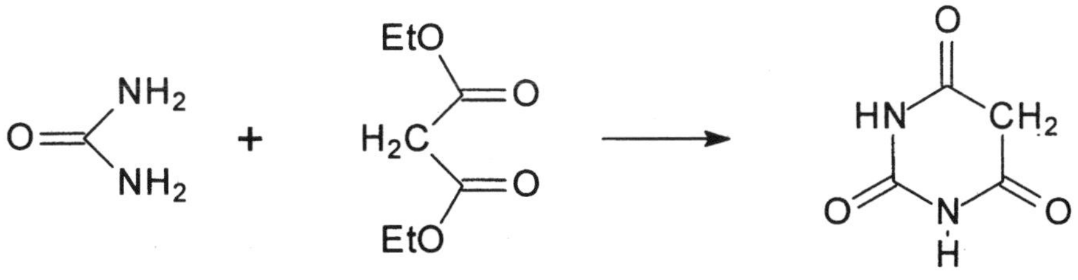

urea + diethylmalonate → barbituric acid

The general mechanism of these acyl substitution reactions involves attack of the entering nucleophile (Lewis base) on the electron-deficient carbonyl carbon (Lewis acid) to initially produce the tetrahedral intermediate structure shown in brackets. This then collapses to reform the carbonyl with expulsion of the leaving group, L^-.

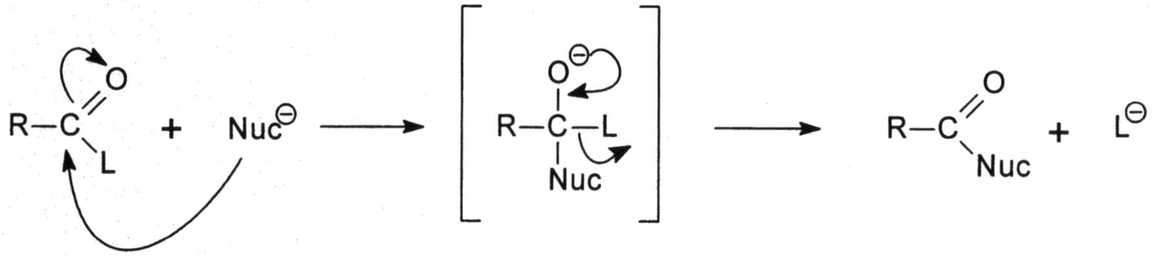

These reactions are often catalyzed by either acid or base. For example, in ester hydrolysis:

Acid Catalyzed Mechanism

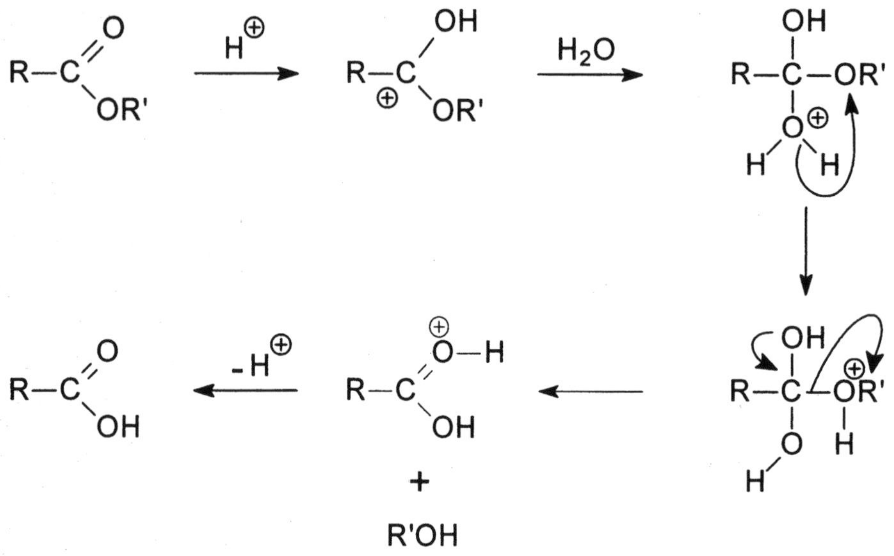

Base Catalyzed Mechanism

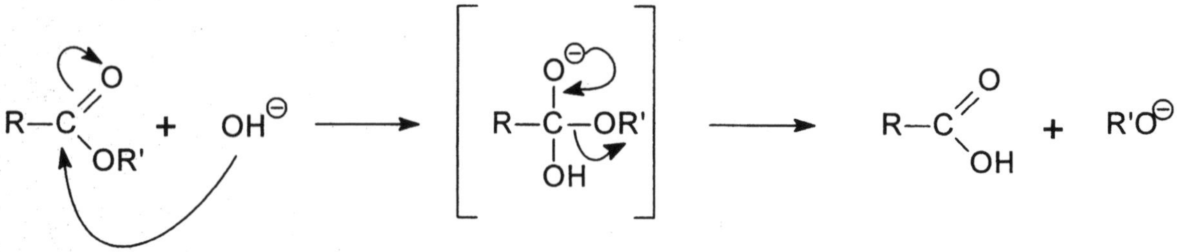

Protonation of the carbonyl oxygen by an *acid* catalyst makes the carbonyl carbon more strongly positive, and for this reason more attractive to nucleophilic attack by water, thus increasing reaction rate. *Base* catalyzes the substitution reaction since the hydroxide ion is a stronger nucleophile than water. We will see many similar examples of acid and base catalysis in organic chemistry.

Active Esters

A clever strategy is often employed to enhance the rate and ease of converting carboxylic acids into their various derivatives. This involves a two-step reaction sequence in which the carboxyl OH group is temporarily replaced by a better leaving group, shown as X below, forming a highly reactive *"active ester"* intermediate. This is then reacted with the nucleophile, Nuc⁻, to rapidly produce the desired substitution product.

| carboxylic acid | "active ester" intermediate | substitution product |

The OH^- ion is the conjugate anion of H_2O, which is a weak acid. It is a poor leaving group for that reason. Therefore, direct substitution on carboxylic acids generally proceeds slowly. The carboxylic acid, however, can be easily converted to an *acid chloride* most often using the reagent thionyl chloride, $SOCl_2$. Now the leaving group is the chloride ion. This is the anion of HCl, a very strong acid, and is thus a very good (fast-reacting) leaving group. Reaction of a nucleophile with the intermediate acid chloride thus leads to the substitution product easily, quickly, and in high yield:

acid chloride

A valuable family of reagents has been developed to produce various "active ester intermediates" for the purpose of effecting rapid and efficient carbonyl substitution reactions. These include NHS (N-hydroxysuccinimide), and carbodiimide reagents, among others. (Unlike acid chlorides, these can be used in water solution, which greatly enhances their usefulness—especially for biochemical applications involving water soluble proteins.)

123

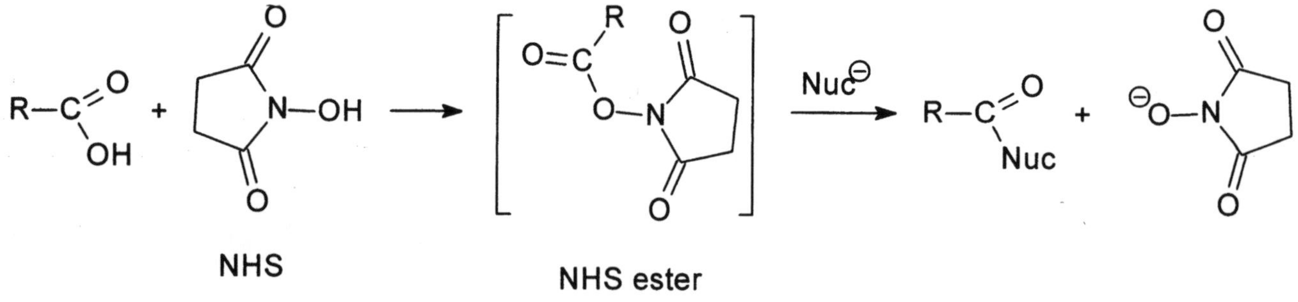

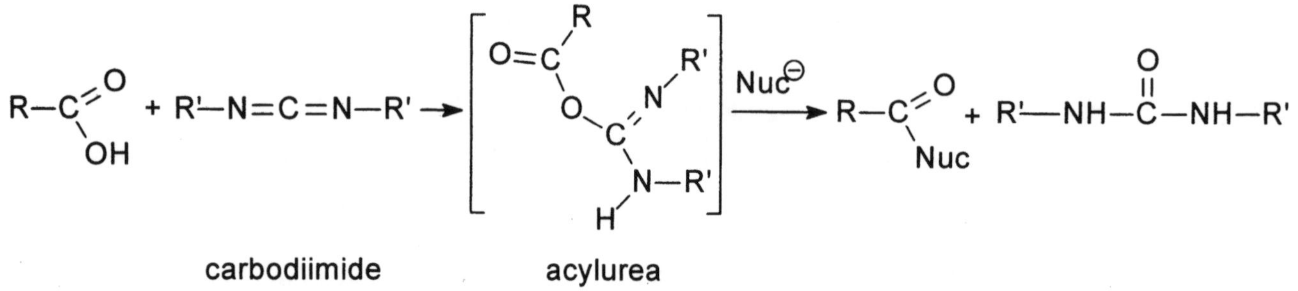

NHS   **NHS ester**

carbodiimide   **acylurea**

Finally, note that this strategy of employing an "active ester" intermediate is entirely analogous to the use of tosyl chloride to enhance the rate of nucleophilic substitution for OH groups at saturated carbon (See Chapter 22).

Polyamides and Proteins

Polyamides such as nylon can be prepared using the reactions described in this chapter. Nylon is a polymer formed by the reaction of a dicarboxylic acid, adipic acid, with a diamine, 1,6-hexanediamine.

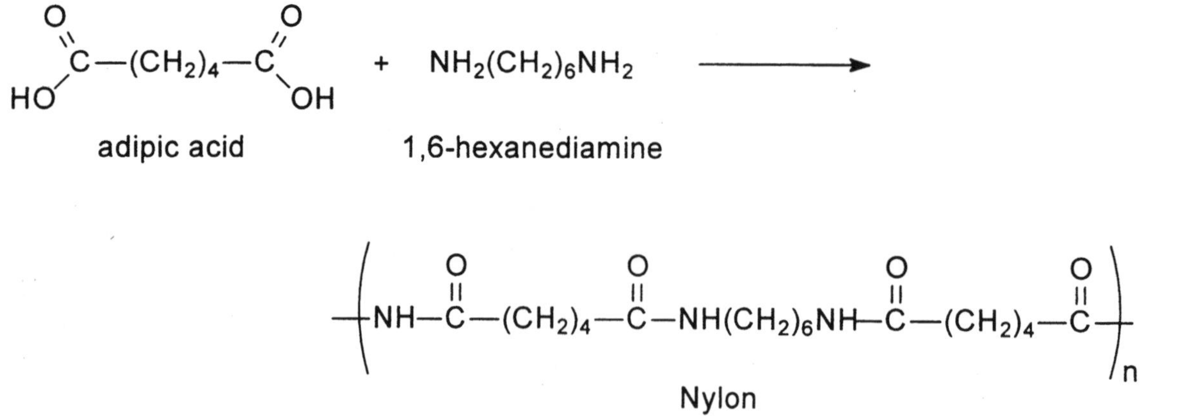

adipic acid   1,6-hexanediamine

Nylon

Notice that both reacting molecules must be *bi-functional*. That is, they must contain *two* reactive functional groups. This condition is necessary for this type of polymerization reaction, since reactive carboxyl and amino groups must always be present at the ends of the growing molecular chain for the reaction to continue.

$$HO-\overset{\overset{O}{\|}}{C}-(CH_2)_4-\overset{\overset{O}{\|}}{C}-OH \quad + \quad NH_2(CH_2)_6NH_2$$

$$\downarrow$$

$$HO-\overset{\overset{O}{\|}}{C}-(CH_2)_4-\overset{\overset{O}{\|}}{C}-NH(CH_2)_6NH_2$$

Another adipic acid and
1,6-hexanediamine

$$NH_2(CH_2)_6NH-\overset{\overset{O}{\|}}{C}-(CH_2)_4-\overset{\overset{O}{\|}}{C}-NH(CH_2)_6NH-\overset{\overset{O}{\|}}{C}-(CH_2)_4-\overset{\overset{O}{\|}}{C}-OH$$

Another adipic acid and
1,6-hexanediamine

etc.

Polyamide formation involving α-amino carboxylic acids (amino acids) is arguably the most important general reaction in the biological world. The products of these reactions are called *polypeptides* or *proteins* and include the enzymes, hormones, structural and transport molecules that define and maintain livings cells.

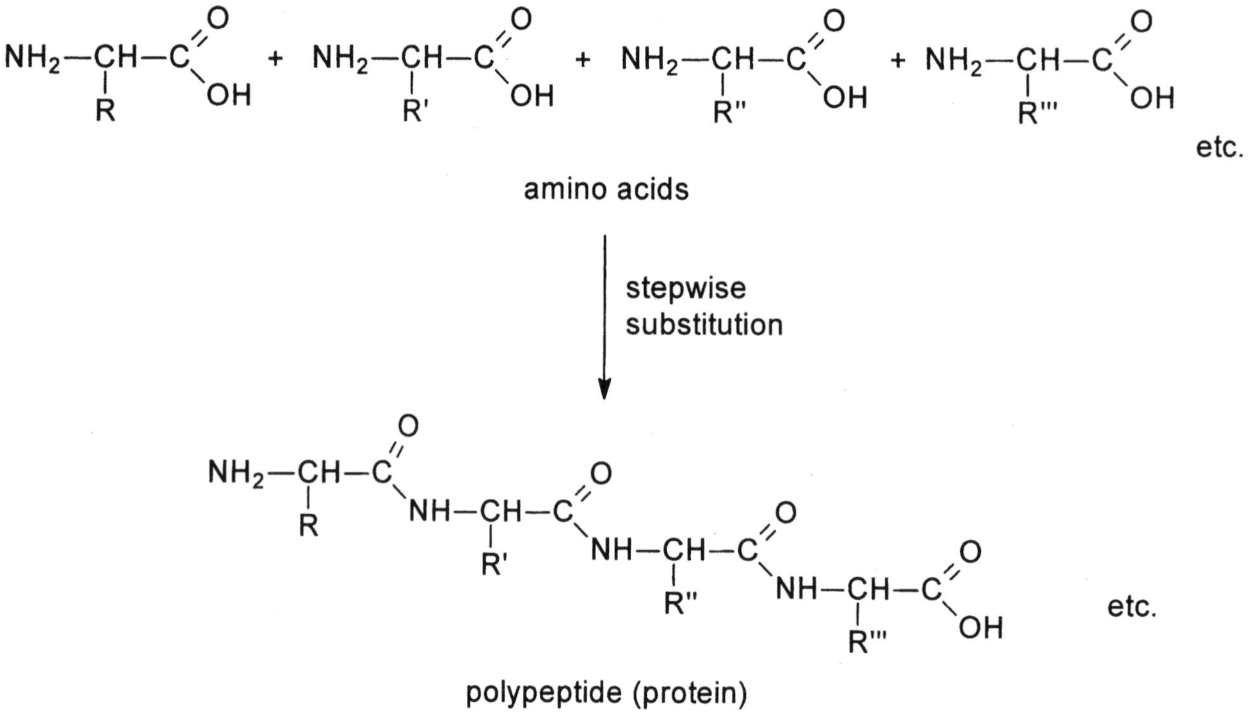

$$NH_2—CH—C{\overset{O}{\underset{OH}{}}} \quad + \quad NH_2—CH—C{\overset{O}{\underset{OH}{}}} \quad + \quad NH_2—CH—C{\overset{O}{\underset{OH}{}}} \quad + \quad NH_2—CH—C{\overset{O}{\underset{OH}{}}} \quad \text{etc.}$$

$$\underset{R}{} \qquad \underset{R'}{} \qquad \underset{R''}{} \qquad \underset{R'''}{}$$

amino acids

stepwise
substitution

polypeptide (protein)

There are about 20 such naturally occurring amino acids, differing only in the structure of the side chain on the alpha- carbon atom, shown as R′, R″, R‴ etc., in the equation above. Amino acid polymerization in a living cell occurs with exquisite specificity, incorporating the various amino acid monomers in an exact sequence to form a polymeric protein uniquely suited for a specific biological function. (Interestingly, the biochemical mechanism of this polymerization involves the conversion of the amino acid carboxyl groups into *acyl phosphate ester* derivatives preceding amide bond formation. These phosphate ester intermediates, like the acid chlorides and other activating groups described above, offer a *better leaving group* than the OH⁻ of the carboxylic acid, thus permitting the biochemical reaction to proceed quickly and efficiently.)

Reduction

Carboxylic acids are reduced to the corresponding 1° alcohol by lithium aluminum hydride, $LiAlH_4$, a very strong reducing agent. The aldehyde which is initially formed is rapidly reduced further to the alcohol by this reagent. Ester or acid chloride derivatives of carboxylic acids, and even carboxylic acids themselves, can also be reduced to alcohols in this way. More selective hydride-donating reagents have been developed which have the ability to produce high yields of the intermediate aldehyde since they do not further reduce the aldehyde to the alcohol. These reagents, shown below, include DIBAL-H and tri-t-butoxyaluminum hydride. Their selectivity is related to steric hindrance of the bulky t-butyl or sec-butyl groups which inhibits close approach of the hydride to the carbonyl carbon, moderating reactivity and preventing further reduction.

Lithium Aluminum Hydride

$$R-C\begin{array}{c}O\\\\OH\end{array} \quad \xrightarrow[\text{2) } H_2O]{\text{1) LiAlH}_4} \quad R-CH_2OH$$

Diisobutylaluminum Hydride (DIBAL-H)

$$R-C\begin{array}{c}O\\\\OR'\end{array} \quad \xrightarrow[\text{2) } H_2O]{\text{1) HAl(i-butyl)}_2} \quad R-C\begin{array}{c}O\\\\H\end{array}$$

Lithium Tri-*tert*-butoxyaluminum Hydride

$$R-C\begin{array}{c}O\\\\Cl\end{array} \quad \xrightarrow[\text{2) } H_2O]{\text{1) Li}^{\oplus}\left[HAl\left(O-\overset{CH_3}{\underset{CH_3}{\overset{|}{\underset{|}{C}}}}-CH_3\right)_3\right]^{\ominus}} \quad R-C\begin{array}{c}O\\\\H\end{array}$$

Chapter 26
Reactions of the Carbonyl Group: A Summary

Almost all reactions of the carbonyl group are initiated by nucleophilic attack of a Lewis base on the electron-deficient carbonyl carbon atom.

$$Nuc^{\ominus} \quad \diagup\!\!\!\!C\!=\!O \quad \longrightarrow \quad Nuc\!-\!\overset{|}{\underset{|}{C}}\!-\!O^{\ominus}$$

The oxy-anion initially produced will then further react by one of only a few possible pathways:

1. The oxy-anion can simply pick up a proton from the reaction solvent or during subsequent product isolation:

$$Nuc\!-\!\overset{|}{\underset{|}{C}}\!-\!O^{\ominus} \quad \xrightarrow{H^{\oplus}} \quad Nuc\!-\!\overset{|}{\underset{|}{C}}\!-\!OH$$

Here are a few examples:

Nucleophile	Aldehyde or Ketone		Product

$$H^{\ominus} + \diagup\!\!\!\!C\!=\!O \longrightarrow H\!-\!\overset{|}{\underset{|}{C}}\!-\!O^{\ominus} \xrightarrow{H^{\oplus}} H\!-\!\overset{|}{\underset{|}{C}}\!-\!OH$$

hydride ion   alcohol

$$R^{\ominus} + \diagup\!\!\!\!C\!=\!O \longrightarrow R\!-\!\overset{|}{\underset{|}{C}}\!-\!O^{\ominus} \xrightarrow{H^{\oplus}} R\!-\!\overset{|}{\underset{|}{C}}\!-\!OH$$

carbanion   alkylated alcohol

$$H_2O + \diagup\!\!\!\!C\!=\!O \longrightarrow H\overset{\oplus}{\underset{H}{O}}\!-\!\overset{|}{\underset{|}{C}}\!-\!O^{\ominus} \longrightarrow HO\!-\!\overset{|}{\underset{|}{C}}\!-\!OH$$

(water or OH$^{\ominus}$)   hydrate

$$ROH + \diagup\!\!\!\!C\!=\!O \longrightarrow R\overset{\oplus}{\underset{H}{O}}\!-\!\overset{|}{\underset{|}{C}}\!-\!O^{\ominus} \longrightarrow RO\!-\!\overset{|}{\underset{|}{C}}\!-\!OH$$

(alcohol or $^{\ominus}$OR)   hemiacetal

$$CN^{\ominus} + \diagup\!\!\!\!C\!=\!O \longrightarrow N\!\equiv\!C\!-\!\overset{|}{\underset{|}{C}}\!-\!O^{\ominus} \xrightarrow{H^{\oplus}} N\!\equiv\!C\!-\!\overset{|}{\underset{|}{C}}\!-\!OH$$

cyanide   cyanohydrin

2. In some cases the initially formed product can undergo dehydration (loss of water) under the conditions of the reaction. This will happen if the attacking nucleophile bears a *hydrogen,* which can be easily lost as a *proton* along with the *hydroxide* derived from the carbonyl oxygen. This is typically the case when the nucleophile is an amine or amine derivative. This sequence of *addition* followed by *elimination* results in *substitution.*

A few examples of this pathway follow:

Nucleophile	Aldehyde or Ketone		Product

R—NH₂ amine; C=O; proton transfer; R—NH—C—OH; -H₂O; R—N=C Schiff base

HONH₂ hydroxylamine; C=O; proton transfer; HO—NH—C—OH; -H₂O; HO—N=C oxime

NH₂NH₂ hydrazine; C=O; proton transfer; NH₂—NH—C—OH; -H₂O; NH₂—N=C hydrazone

ArNHNH₂ arylhydrazine; C=O; proton transfer; ArNH—NH—C—OH; -H₂O; ArNH—N=C arylhydrazone

3. If the carbonyl compound is a *carboxylic acid or carboxylic acid derivative,* it will contain a leaving group (L) which can be lost upon reformation of the carbonyl bond from the oxy-anion in brackets. (This is another example of an addition-elimination pathway resulting in substitution.)

129

Examples of this pathway include all the substitution reactions of carboxylic acids and derivatives. For example:

Ester Hydrolysis

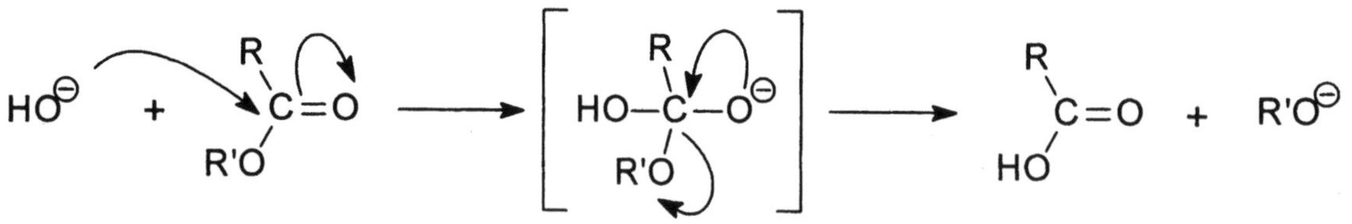

$$HO^{\ominus} + \underset{R'O}{\overset{R}{C}}=O \longrightarrow \left[\underset{R'O}{\overset{R}{HO-C-O^{\ominus}}} \right] \longrightarrow \underset{HO}{\overset{R}{C}}=O + R'O^{\ominus}$$

Amide Synthesis

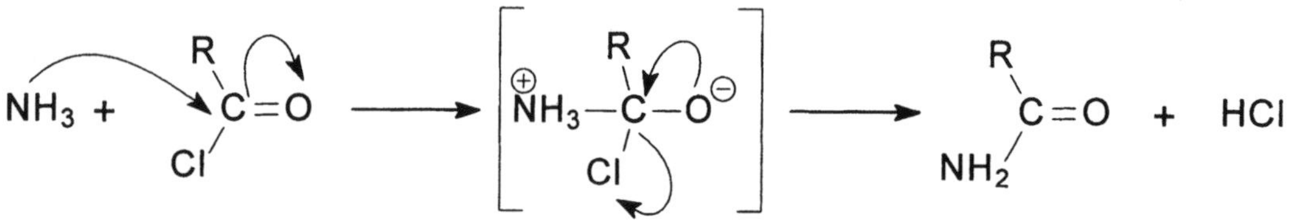

$$NH_3 + \underset{Cl}{\overset{R}{C}}=O \longrightarrow \left[\underset{Cl}{\overset{R}{\overset{\oplus}{N}H_3-C-O^{\ominus}}} \right] \longrightarrow \underset{NH_2}{\overset{R}{C}}=O + HCl$$

Transesterification

$$R''OH + \underset{R'O}{\overset{R}{C}}=O \longrightarrow \left[\underset{\overset{\oplus}{R'O}}{\overset{H\ \ R}{R''O-C-O^{\ominus}}} \right] \longrightarrow \underset{R''O}{\overset{R}{C}}=O + R'OH$$

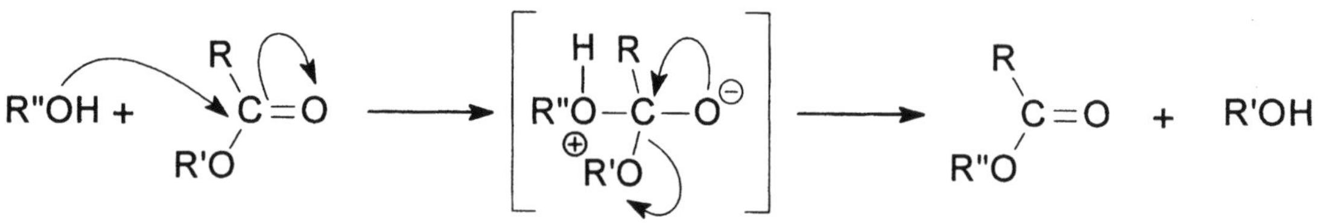

Carbonyl addition and substitution reactions are an excellent example of how a great variety of reactions in organic chemistry proceed by a similar mechanism. *The initial step in each case is the attack of a nucleophile on the electron-deficient carbonyl carbon atom.* Subsequent steps vary, depending on the specific structure of the reacting carbonyl compound and nucleophile.

Chapter 27
Resonance Structures and Electron Delocalization

Consider an unpaired electron on a carbon atom adjacent to a double bond. This is called an *allyl radical:*

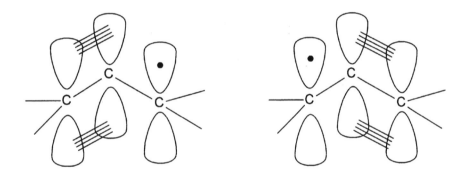

The pi electron cloud of the alkene and the orbital containing the unpaired electron overlap above and below the sigma bond. This is a system of three adjacent carbon atoms, each contributing one electron in a dumbbell shaped p orbital overlapping one or both of its neighbors. The likelihood of pi bond formation to the right or left has equal probability for these two mirror image structures, called *resonance structures.*

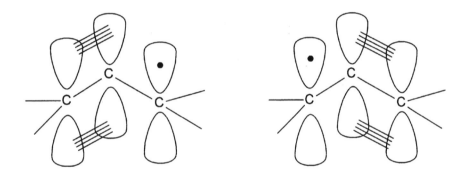

The allyl radical can thus be better described by equal contributions of the two resonance structures than by either structure alone. *The weighted average of these resonance structures describes the electron distribution more accurately than any individual structure.* The p orbital electrons in the allyl radical are said to be *delocalized,* since they are distributed among several bonds.

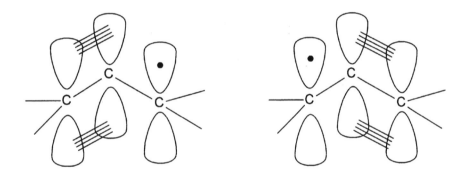

equivalent to

The double-headed arrow implies that the electronic distribution in this radical is described by a *simultaneous* superimposition of these two structures. Because of the symmetry of this radical, each structure contributes an equal fraction (50%) to the overall description of the allyl radical, with each of the two end carbon atoms carrying 1/2 of the unpaired electron. The dotted lines indicate partial (1/2) bonds resulting from the average of the two resonance structures. (The average of a double bond and a single bond is 1 1/2 bonds linking each terminal carbon to the central carbon.)

Electron delocalization confers special stabilization on this radical. Since the unpaired electron is *shared* between two atoms, no *single* atom has to carry the unstable electron septet alone. The delocalization of p orbital electrons described by these resonance structures lowers energy and enhances stability. This happens whenever overlap occurs between at least three adjacent p orbitals—in reactive intermediates, including free radicals, carbocations, and carbanions; in poly-atomic ions; and in neutral molecules.

The nitrate, NO_3^- and carbonate, CO_3^{-2} ions are familiar examples from inorganic chemistry. In both cases, orbital overlap permits the equal contribution of each of three resonance structures. Each of these makes a 1/3 contribution to the overall description of these ions, so that on average, 1 1/3 bonds link the central atom to each oxygen, and each oxygen carries a charge of $-2/3$.

Nitrate, $NO_3^{\ominus}$

Carbonate, CO_3^{-2}

Resonance structures play an important role in organic chemistry since they describe the electron distribution in a molecule, ion, or reactive intermediate more accurately than any individual structure could. In the following chapters, we shall see that consideration of resonance structures permits us to make accurate predictions about reactivity rates and product isomer distributions, as well as other chemical and physical properties of organic compounds. The key facts are these:

1. Resonance requires p orbital overlap between adjacent atoms.
2. Resonance structures describe the delocalization of p orbital electrons.
3. Resonance delocalization increases the stability of a covalent molecular system relative to any individual localized structure.

4. Reaction rates increase when reactive intermediates can be described by several reasonance structures, since the more stable intermediate will form with greater ease.

5. High or low yields of various products in a reaction mixture can often be understood by considering the number and stability of resonance structures describing the reactive intermediate leading to each product. The more stable intermediate forms faster and gives rise to the major product.

The following examples show how resonance structures describe neutral molecules and various reactive intermediates:

Naphthalene

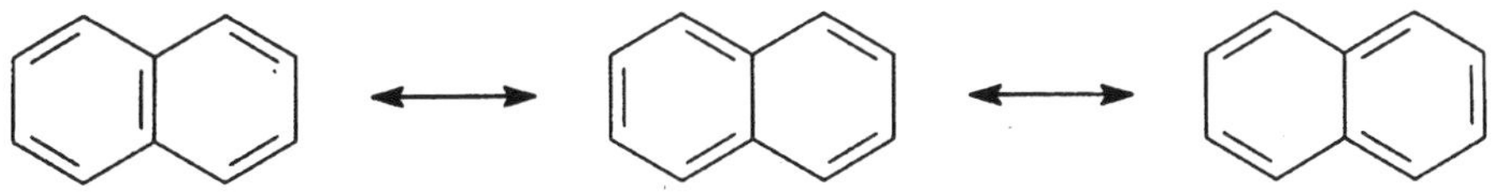

Galvinoxyl (a stable free radical)

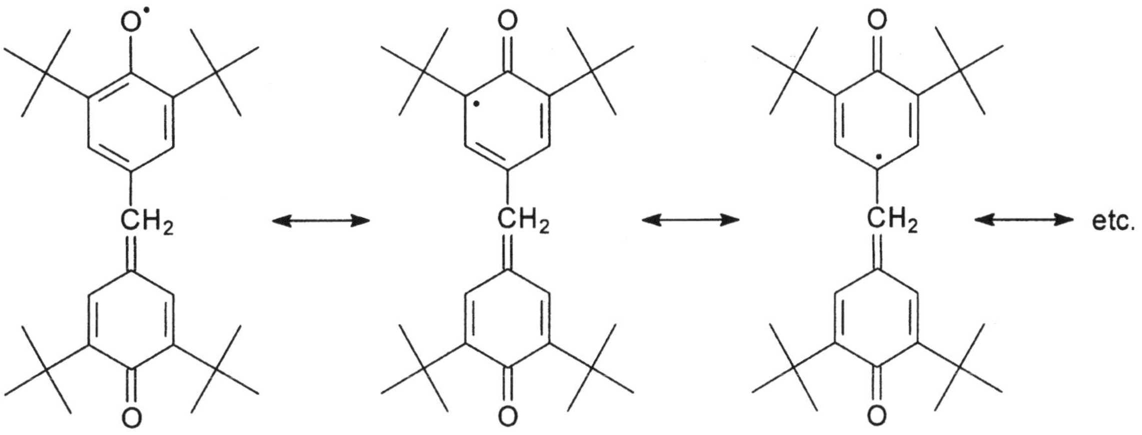

The Benzyl Carbocation

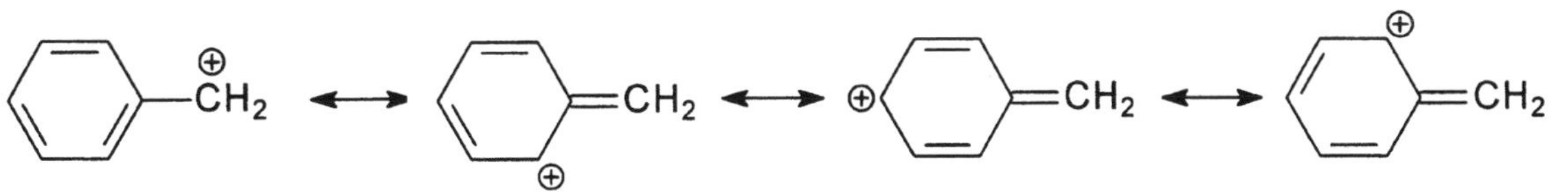

Malonic Ester Carbanion

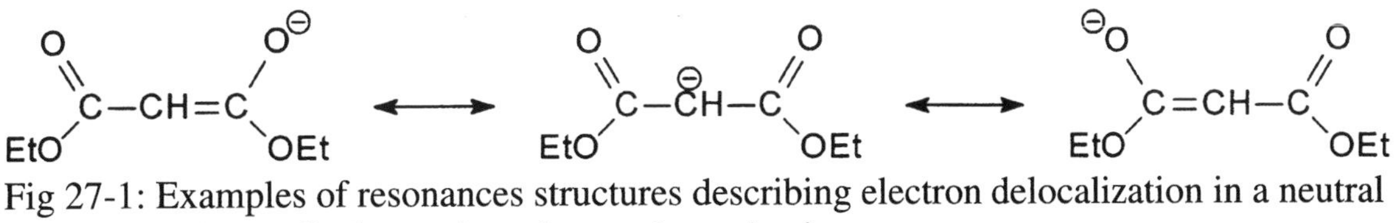

Fig 27-1: Examples of resonances structures describing electron delocalization in a neutral molecule; a free radical; a carbocation; and a carbanion.

In each case, resonance structures are written by moving either single p orbital electrons or electron pairs to form bonds with a p orbital *on an adjacent atom*. This process is often referred to as"electron flipping" and is often indicated by curved arrows. Note that in these examples, this "flipping" procedure results in the transfer of positive or negative charge or an unpaired electron onto *alternate* atoms and places unsaturation on *adjacent* bonds. The following examples show the use of curved arrows to derive resonance structures for several delocalized systems. (The resonance structures shown in Fig 27-1 were derived in exactly this manner.)

Neutral Molecule

Free Radical

Carbanion

$$\overset{\ominus}{C}H_2 - \overset{\overset{\textstyle O}{\|}}{C} - CH_3 \longleftrightarrow CH_2 = \overset{\overset{\textstyle O^{\ominus}}{|}}{C} - CH_3$$

Carbocation

Fig 27-2: Resonance structures derived by "electron flipping".

Chapter 28
Aromatic Compounds and Aromaticity

Consider the molecule *benzene:* C_6H_6. This compound is composed of 6 C-C sigma bonds forming a hexagon and 6 C-H sigma bonds, as well as three pi bonds which result from the overlap of p orbitals on adjacent carbon atoms. Overlap of p orbitals can occur with equal probability with a neighboring p orbital to the left (counterclockwise) or to the right (clockwise):

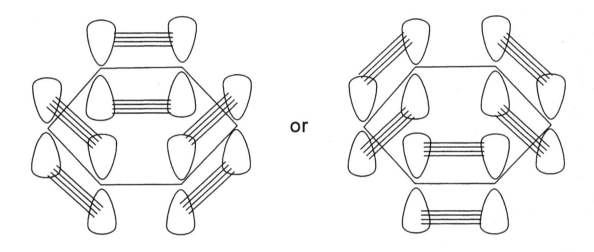

For this reason the pi bonds of the benzene molecule are described by equal *contributions of the two* resonance structures, to the right and left of the double-headed arrow. The average of these is sometimes written as a hexagon with an inner dashed circle.

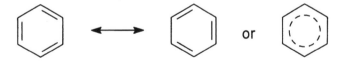

The dashed line stands for 1/2 of a pi bond, resulting from the average of the contributions of each of the two resonance structures. These structures are *not* in equilibrium with each other; rather, the molecule is understood to be a *simultaneous superimposition* of equal contributions of the two structures. (For convenience, in routine work, only one of the resonance structures is typically written; but the other is always implied. Also, a *full* circle within a hexagon is a shorthand often used to describe benzene and its derivatives.)

Certain molecules like benzene are said to be *aromatic. Aromatic compounds possess a special stabilization due to electron delocalization completely around a cyclic system.* This stabilization can be demonstrated experimentally by conducting the exothermic catalytic hydrogenation (reduction) of benzene along with the model compound cyclohexene. The heat evolved in these reactions can be measured in a calorimeter. The heat evolved in the hydrogenation of *one* mole of benzene is found to be *36 kcal/mole less* than is evolved when *three* moles of cyclohexene are reduced in the same way. This 36 kcal/mole difference is due to resonance delocalization of the pi electrons and is called the *aromatic stabilization energy* of benzene. Aromatic stabilization significantly lowers the energy content of a molecule and so affects its chemical reactivity.

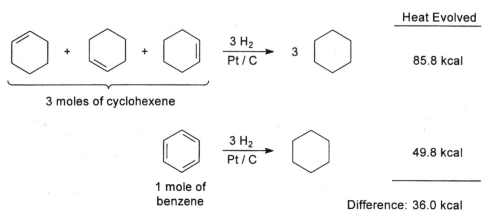

Fig 28-1: Demonstration of the aromatic stabilization energy of benzene by calorimetry.

To realize the benefit of aromatic stabilization, a molecule must have the following characteristics:

1. It must have a ring structure in which p orbital overlap can occur *without any discontinuities* all around the ring:

2. The total number of conjugated (delocalized) p orbital electrons must equal $4n + 2$ where n is 0 or a positive integral number. (This is a result of a branch of physics called *quantum mechanics* and is known as the *Hückle Rule.*)

n	4n+2
0	2
1	6
2	10
3	14
4	18
--	--
etc.	etc.

A ring system with 2, 6, 10, 14, 18, . . . overlapping p orbital electrons may be aromatic, and possess special stability for that reason. (Thus benzene, with 6 overlapping p orbital electrons would be aromatic.) Rings with 4, 8, 12, 16, . . . pi electrons will not be stabilized; indeed, theory says that these rings, called *anti-aromatic,* will have a special *instability,* and may be highly reactive or even incapable of existence for that reason. (However, certain ring systems which *do* satisfy the Huckle Rule are still not aromatic due to geometric constraints inhibiting p orbital overlap. See below.)

3. A molecule must be planar, or very nearly so, to permit the p orbital overlap required for aromaticity. In such molecules as the 10 carbon ring system [10]-annulene, planarity is not possible due to steric hindrance between the interior H atoms. This forces the p-orbitals of these molecules out of co-planarity, thus inhibiting the required p-orbital overlap:

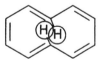

[10]-annulene

Other categories of aromatic compounds also exist. Multi-ring systems are also capable of aromatic stabilization, as are molecules in which some of the contributing electrons are donated by the non-bonded electron pairs of heteroatoms such as nitrogen, sulfur, or oxygen. (Rehybridization of an sp³ non-bonded electron pair to a p orbital occurs in these hetero-atomic aromatics to optimize overlap with adjacent p or-bitals.) Thus, the following molecules are also aromatic:

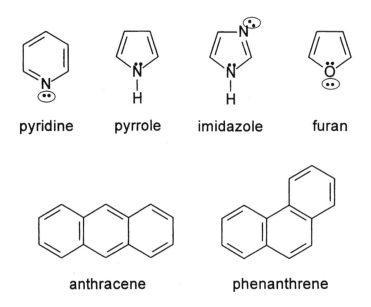

pyridine   pyrrole   imidazole   furan

anthracene   phenanthrene

(Note: The *circled* electron pairs above are *not* involved in the aromatic pi electron cloud. They reside in hybrid orbitals, whose shape is perpendicular to the overlapping p orbitals of the ring. They cannot par-ticipate in the cyclic aromatic "ring current" for this reason.)

The larger "annulenes" either display aromatic properties or do not, depending on the application of the Hückle Rule to each:

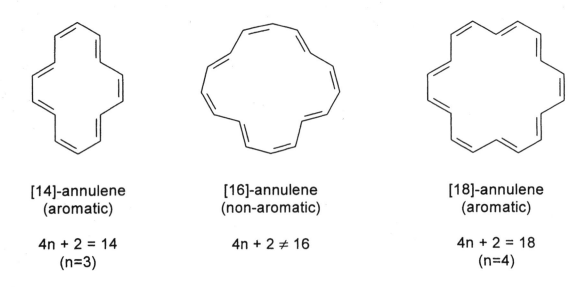

[14]-annulene
(aromatic)

$4n + 2 = 14$
$(n=3)$

[16]-annulene
(non-aromatic)

$4n + 2 \neq 16$

[18]-annulene
(aromatic)

$4n + 2 = 18$
$(n=4)$

Aromatic stabilization is also possible for reactive intermediates, giving them special stability and ease of formation compared to similar non-aromatic carbanions or carbocations.

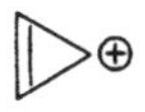

cyclopropenium
cation
$4n + 2 = 2$
$(n = 0)$

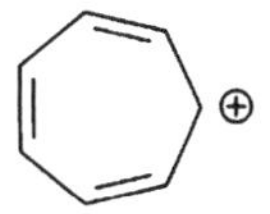

tropylium ion
(cycloheptatrienyl cation)
$4n + 2 = 6$
$(n = 1)$

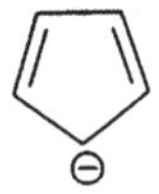

cyclopentadienyl
anion
$4n + 2 = 6$
$(n = 1)$

(Note that in the two carbocations above, the *empty* p orbital at the site of the positive charge allows for *continuity of p orbital overlap* around the entire ring, which is mandatory for aromatic stabilization. Note as well that resonance structures can be written delocalizing the positive or negative charge on *all* ring atoms so that each carries an equal share of the charge.)

Resonance structures describing the p orbital electron delocalization in several of these aromatic systems are shown below:

[14]-Annulene

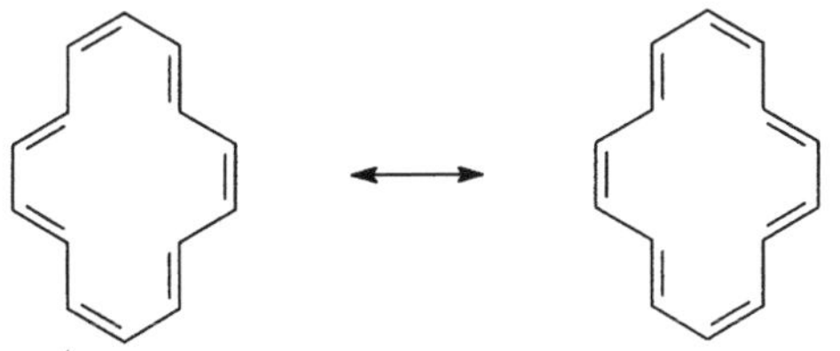

Cyclopentadienyl anion

(Each C atom carries 1/5 of the negative charge.)

Tropylium cation

(Each C atom carries 1/7 of the positive charge.)

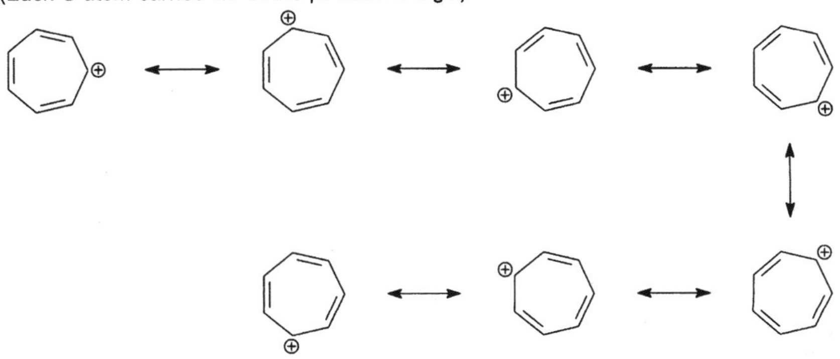

Fig 28-2: Resonance structures in aromatic systems.

Aromaticity and Chemical Reactivity

Aromatic stabilization has a major impact on chemical reactivity. While most unsaturated compounds, like alkenes, undergo *addition* of water, halogen molecules, and mineral acids, these *addition* reactions do not generally occur with aromatic molecules, since the product would no longer be aromatic. By contrast, aromatic compounds react in ways that *preserve* the stable aromatic character of the molecule. They engage in reactions which *substitute* atoms or groups for ring hydrogen atoms, rather than *adding* to pi bonds. To be described as "aromatic," a compound is expected to react in this characteristic manner. In fact, in certain borderline cases, chemical reactivity is used as a criterion to classify a molecule as being aromatic or not.

BIOCHEMISTRY DEPARTMENT

RESTROOMS

Chapter 29
Electrophilic Aromatic Substitution

The most important reaction of aromatic compounds is the substitution of an electron deficient group, E^+, called an *electrophile,* for a ring proton. As a class these reactions are called *Electrophilic Aromatic Substitution.* A wide variety of electrophiles can be employed to produce many different substituted aromatics. The general reaction is represented by the equation:

Mechanism of Electrophilic Aromatic Substitution

General Mechanism of Electrophilic Aromatic Substitution

The electrophile E^+, is an electron-deficient species (Lewis acid). In the first step E^+ accepts an electron pair from the pi cloud of the aromatic ring (Lewis base) to form a covalent bond. This produces an intermediate carbocation called an *arenium ion,* which is resonance stabilized by charge delocalization over three atoms of the benzene ring. Loss of a ring proton then completes the reaction, producing the substitution product:

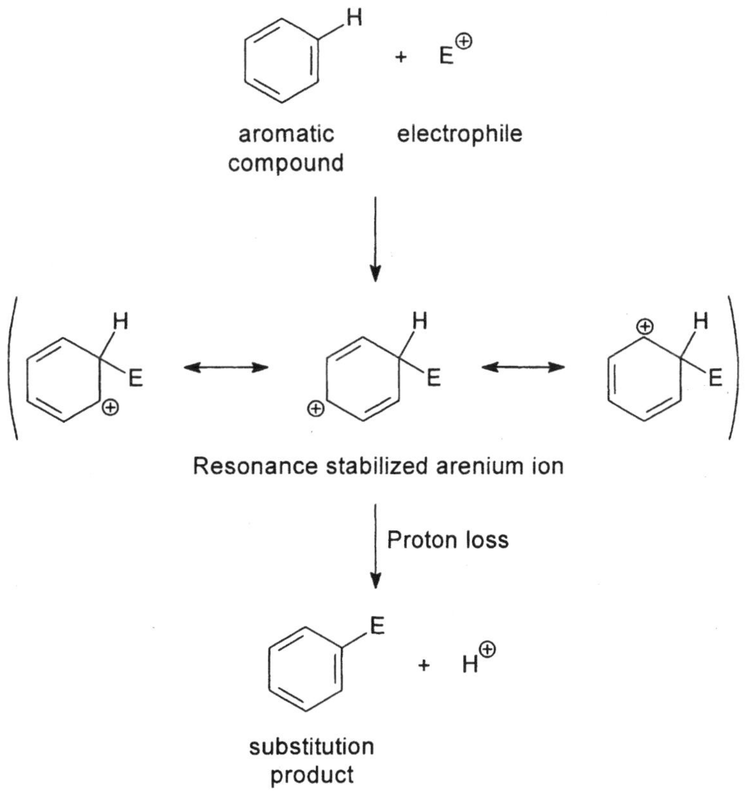

Fig 29-1: Mechanism of electrophilic aromatic substitution.

141

There are five commonly performed reactions of this type. All proceed by the same mechanism; only the nature of the electrophile, E^+, and its method of generation varies from reaction to reaction. These five reactions are summarized in the scheme below:

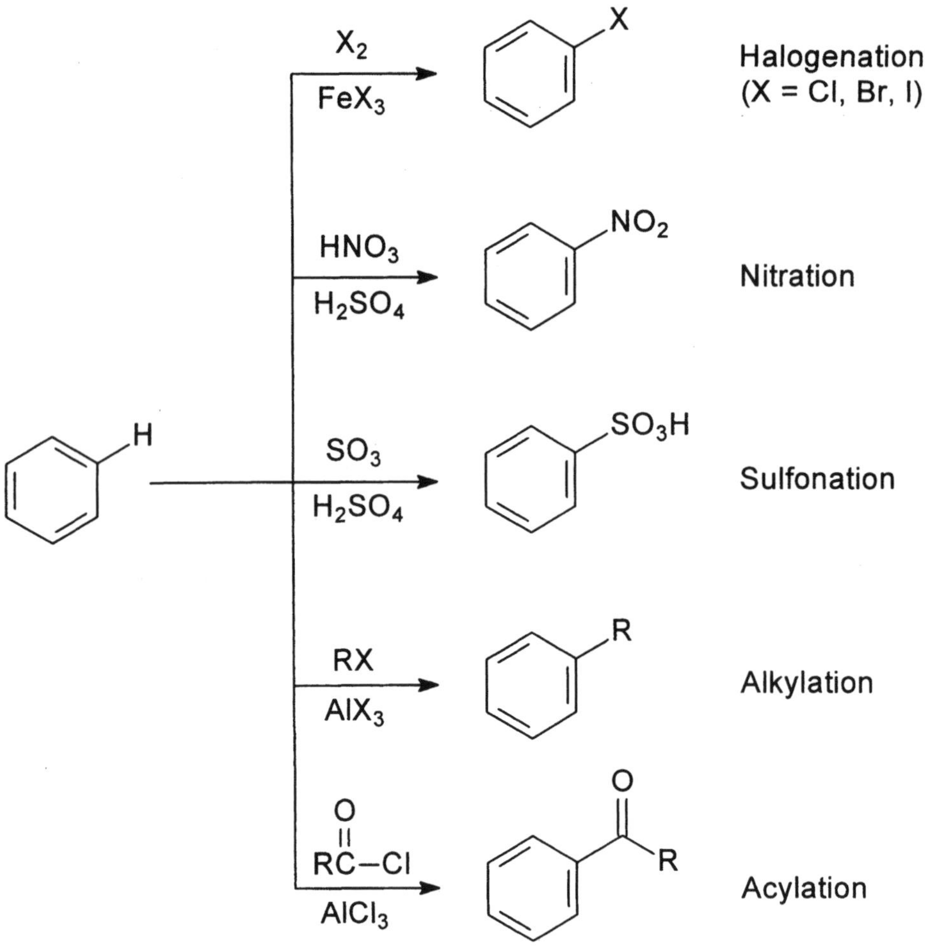

Fig 29-2: The important electrophilic aromatic substitution reactions.

Specific Reaction Mechanisms of Electrophilic Aromatic Substitution

In each of the five reactions in Fig 29-2, reagents interact to produce the electrophile, E^+, shown in brackets in Figs 29-3 through 29-7, below. The electrophile then bonds to the aromatic ring, producing a carbocation intermediate, called an arenium ion. Proton loss then leads directly to the substitution product.

Halogenation
   A catalytic quantity of iron halide interacts with the halogen molecule to generate the electrophilic *halogen cation*, X^+, which reacts with the aromatic ring. The bromine cation is shown below in the preparation of bromobenzene.

142

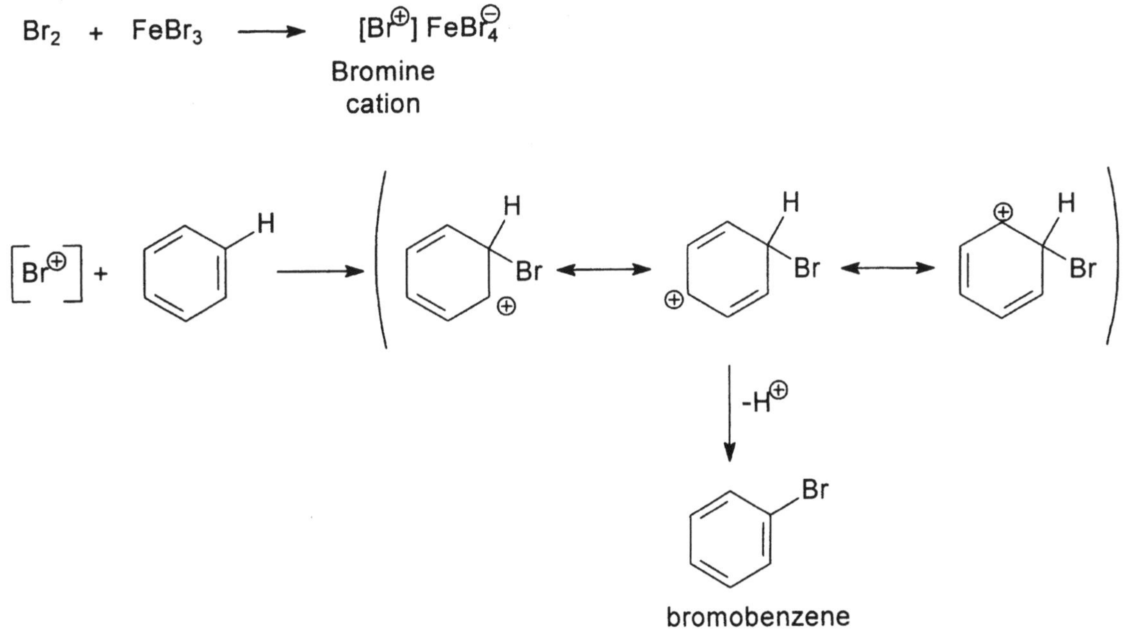

Fig 29-3: Halogenation.

Nitration

The interaction of concentrated sulfuric acid, H_2SO_4, and concentrated nitric acid, HNO_3, produces an equilibrium concentration of an electrophile called the *nitronium ion, NO_2^+*, which then attacks the aromatic ring, forming the nitro-aromatic after proton loss.

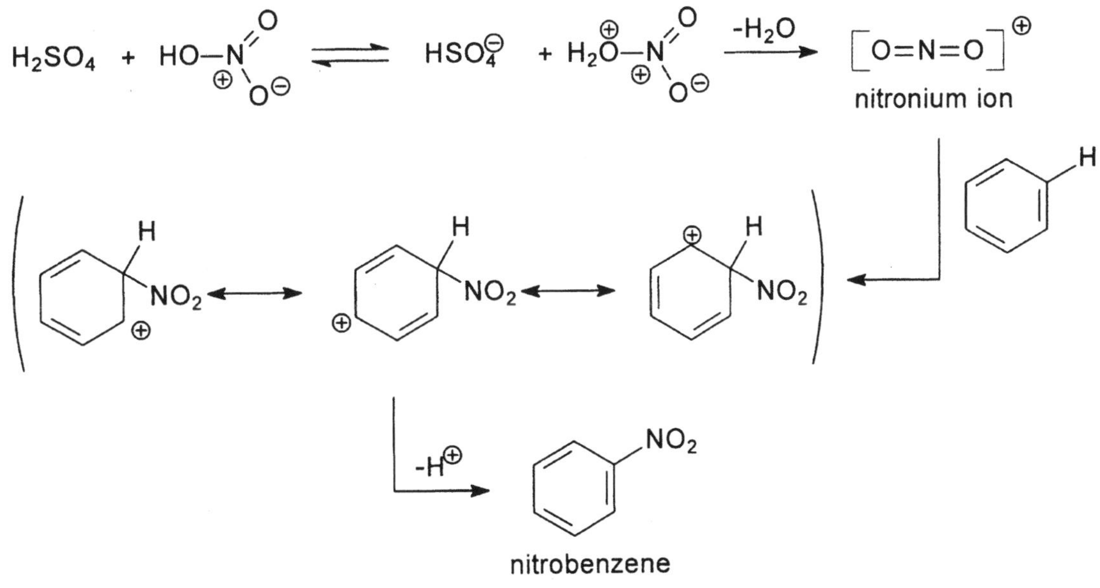

Fig 29-4: Nitration.

143

Sulfonation

In this case, fuming sulfuric acid (excess SO_3 in H_2SO_4) is used as a source of the electrophile SO_3. Attack on the aromatic ring proceeds in accord with the general mechanism to produce an aromatic sulfonic acid.

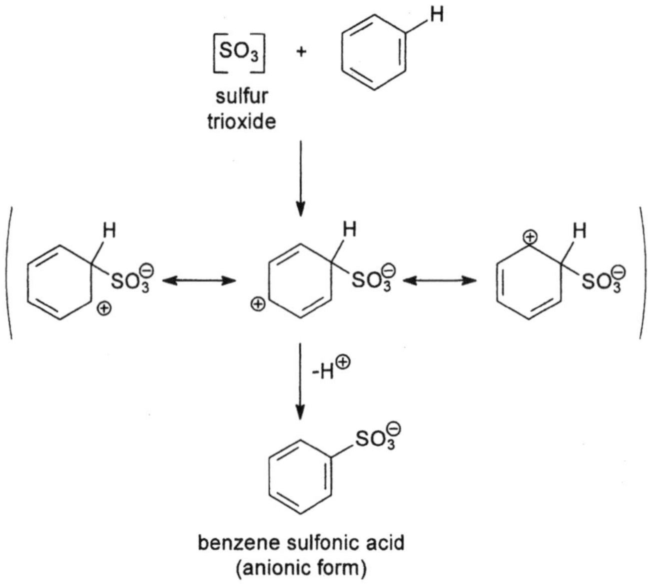

benzene sulfonic acid
(anionic form)

Fig 29-5: Sulfonation.

Alkylation

Here, the electrophile is a carbocation, R^+, produced by reaction of the catalyst, $AlCl_3$, with an alkyl chloride or other alkyl halide. An alkyl substituted aromatic is the product of this reaction.

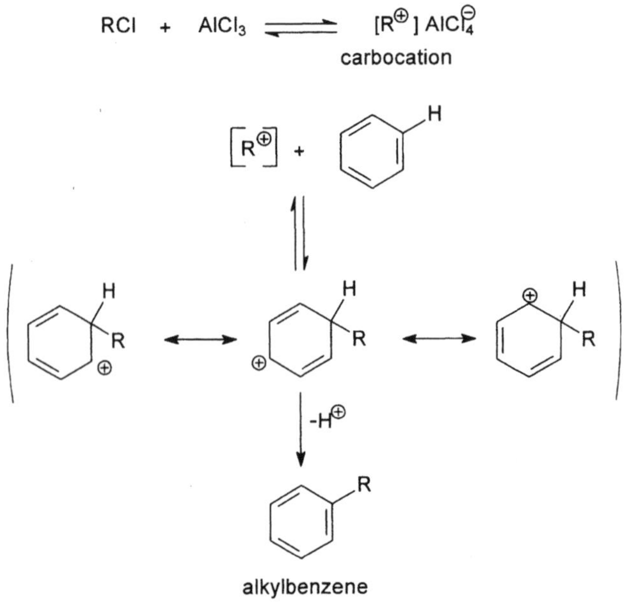

alkylbenzene

Fig 29-6: Alkylation.

Acylation

In acylation, a carboxylic acid derivative with a good leaving group like an acid chloride or anhydride is reacted with the $AlCl_3$ catalyst to generate an electrophilic species called the *acylium ion, RC = O*$^+$. This then proceeds to substitute on the benzene ring as in the previous examples, producing an aromatic ketone as the final product.

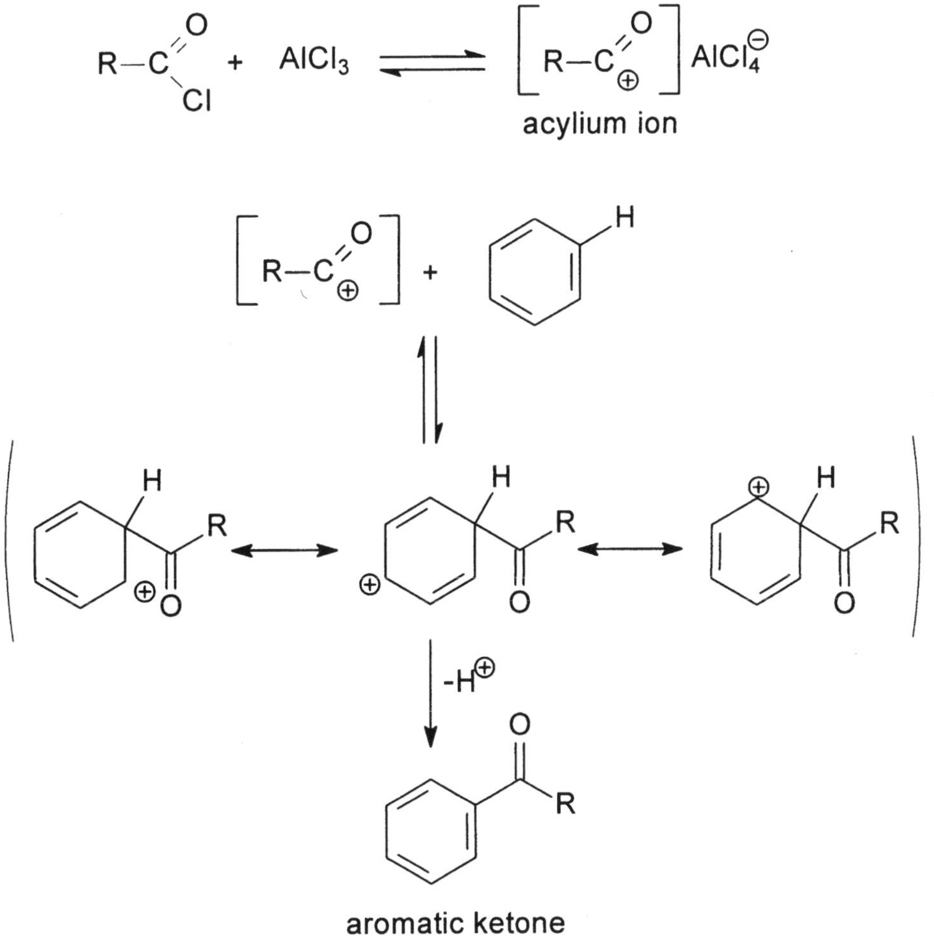

Fig 29-7: Acylation.

The aluminum halide or iron halide catalyst shown in these examples is a strong Lewis acid. Such a catalyst is often needed to encourage formation of the electrophilic species. The catalyst is *electron deficient* since the central atom, Al or Fe, has only *six* valence electrons. It functions by bonding to a halide ion derived from the alkyl halide, acyl halide (acid chloride) or halogen molecule to complete its octet. Simultaneously, the electrophilic carbocation, acylium ion, or halogen cation, is liberated to react with the aromatic compound. Sulfuric acid, H_2SO_4, functions in similar fashion to encourage formation of the nitronium ion, NO_2^+ in nitration reactions. Boron trifluoride can also be used as a catalyst in much the same way. (Some especially reactive aromatic compounds, however, may not require the participation of a Lewis acid catalyst.)

145

Substitution vs Addition Reactions

Electrophiles like Br^+ and H^+ will generally *add* to unsaturated compounds, like alkenes, resulting in addition products by a carbocation mechanism:

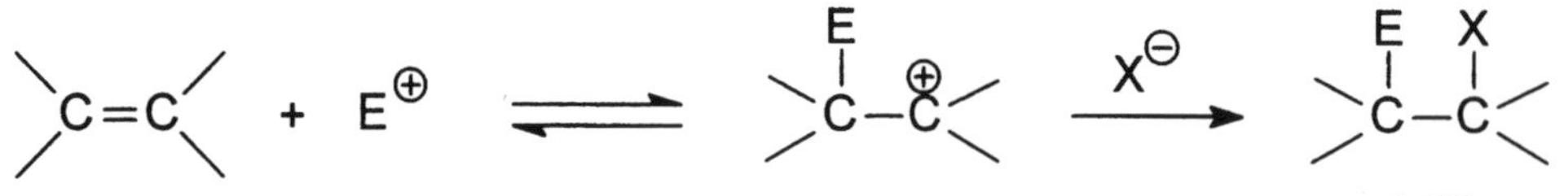

By contrast aromatic compounds tend to undergo *substitution* rather than *addition* because this allows the molecule to preserve its stable aromatic character. Addition reactions would *destroy* aromaticity by creating a discontinuity in the pi electron ring current. *Substitution chemistry is a consequence of the special stability of aromatic compounds.* Participation in ring substitution rather than addition is a hallmark and a defining characteristic of aromatic compounds.

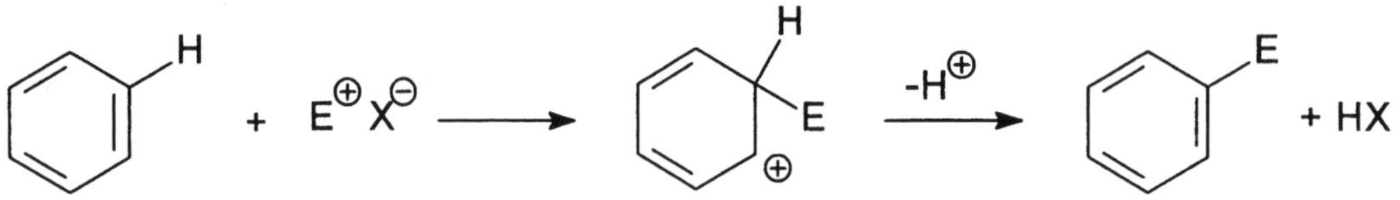

Non-Benzenoid Aromatics

It is not just benzene and its substituted derivatives that are aromatic, but a whole large family of organic compounds that possess planar ring systems having $4n + 2$ p orbital electrons on all adjacent atoms. There exists a wide variety of aromatic compounds defined in this way which engage in the substitution reactions described above. These compounds include fused ring (multi-ring) structures, large monocyclic rings called *annulenes,* and also *heterocycles,* containing ring atoms other than carbon. These reactions proceed by much the same mechanism as benzene itself and produce the same range of halogenated, alkylated, acylated, nitrated, and sulfonated products, and others as well. Here are a few examples of substitution reactions in a selection of these so-called "non-benzenoid" aromatics:

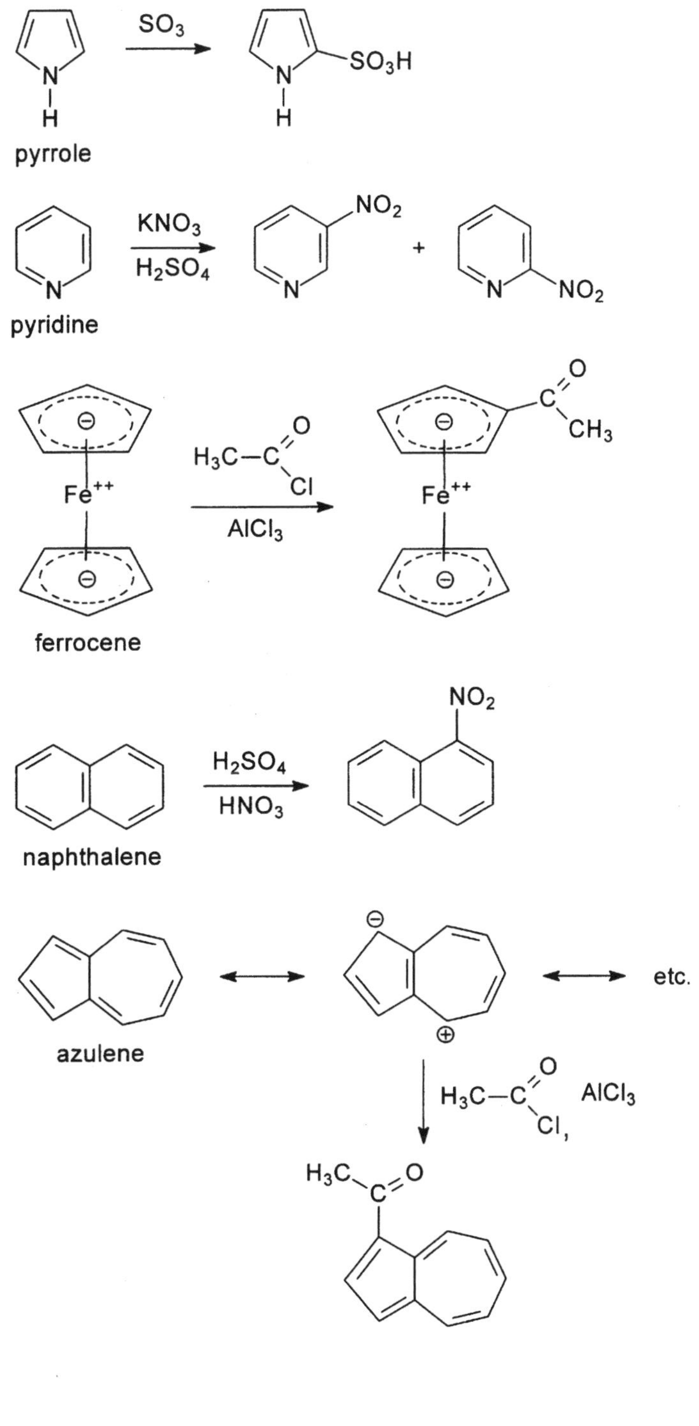

pyrrole

pyridine

ferrocene

naphthalene

azulene

Orientation and Reactivity Effects in Substituted Benzenes

Several isomeric products can be produced when substitution is carried out on a previously substituted ring. Three possible isomers have to be considered. These are the 1,2-(ortho), 1,3-(meta), and 1,4-(para) isomers often abbreviated o, m, and p.

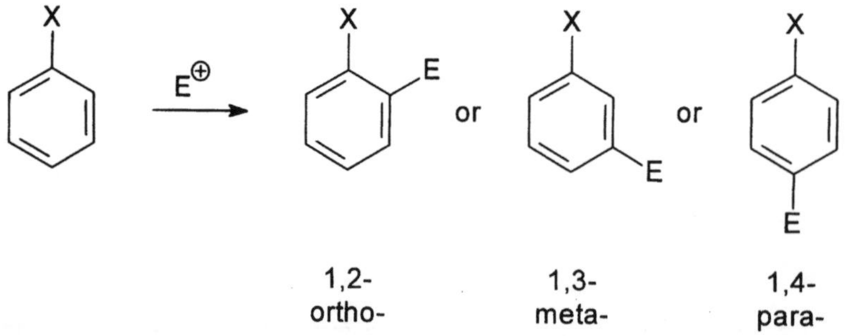

Through long experience it has been found that certain *substituent* groups, shown as X in the preceding illustration, direct further substitution to yield predominantly the o- and p- isomers, while others direct further substitution to mainly yield the m-isomer. Furthermore, the o,p- directors generally *activate* the ring, *increasing* the rate of further substitution. By contrast, the m- directors tend to *deactivate* the ring, *slowing down* the rate of continued substitution. The table below summarizes the various important activators/o,p- directors, on the left, and the deactivators/m- directors on the right:

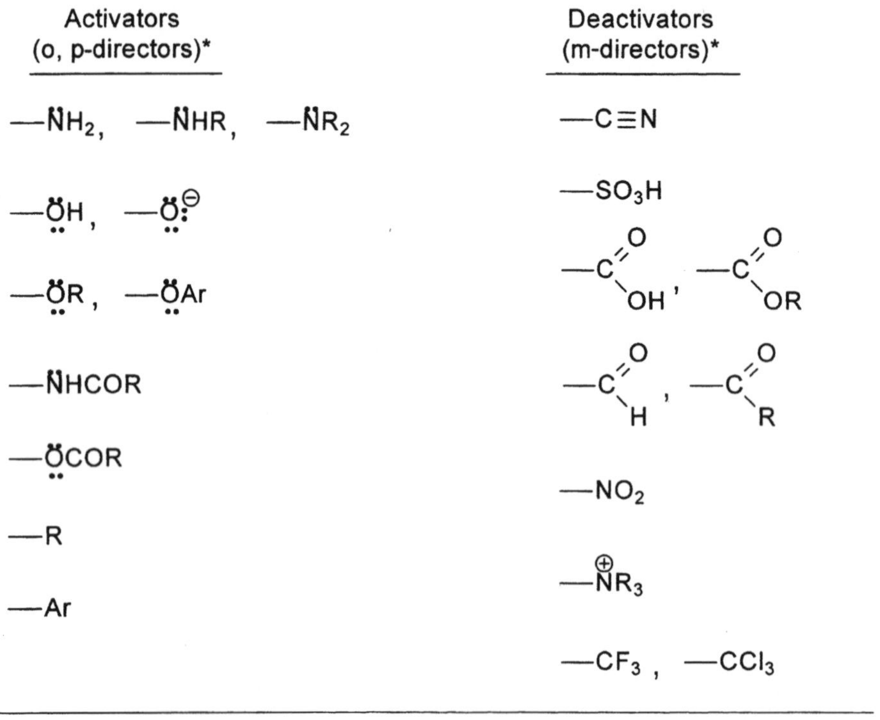

* The halogens (F, Cl, Br, I) are generally o, p directing but deactivating.

Fig 29-8: Substitution effects in electrophilic aromatic substitution.

Resonance structures can be used to explain the trends shown in the preceding table. We'll do this by considering the resonance structures that describe *electron delocalization* in the aromatic reactant and also by considering the resonance structures describing *charge delocalization* in the intermediate arenium ions, leading to the various isomeric products. This is demonstrated below to explain the orienting and reactivity effects of the amino and nitro groups of aniline and nitrobenzene, respectively.

Aniline (o,p-Director and Rate Activator)

The amino group of aniline (aminobenzene) is a strong o,p- director and rate activator. Resonance structures can be written for the aniline molecule in which the non-bonded electron pair on the nitrogen atom interacts with the pi cloud of the aromatic ring:

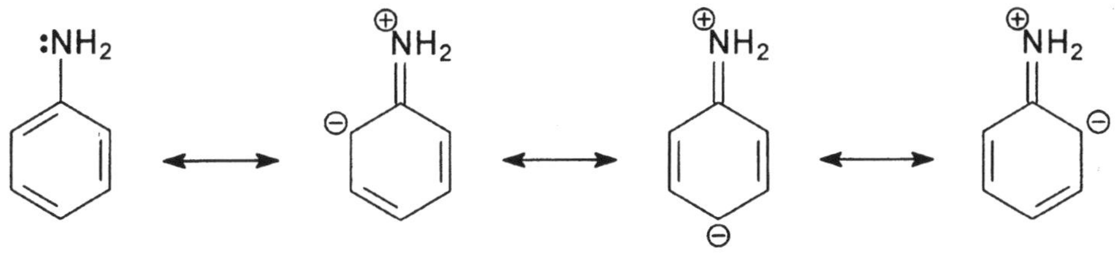

The contribution of the three charged resonance structures results in some degree of extra electron density at the ortho and para carbons, thus predisposing these sites to attack by the electron deficient electrophile.

More important, however, is the fact that the *arenium ion* intermediates resulting from electrophile attack at the o and p carbon atoms are *especially stable*. The underlined resonance structure below, resulting from attack on the ortho carbon, will make a major contribution to stabilizing the cationic intermediate since *it alone has an octet of electrons around all the carbon and nitrogen atoms.* A similar structure stabilizes the carbocation resulting from attack at the para carbon. No such stable structure however, can be written resulting from electrophile attack at the meta carbon atoms. Thus o,p substitution is greatly favored in reactions of aniline, since these isomers result from more highly stabilized arenium ions. The *rate* of ring substitution in aniline is greatly increased compared to benzene for the same reason.

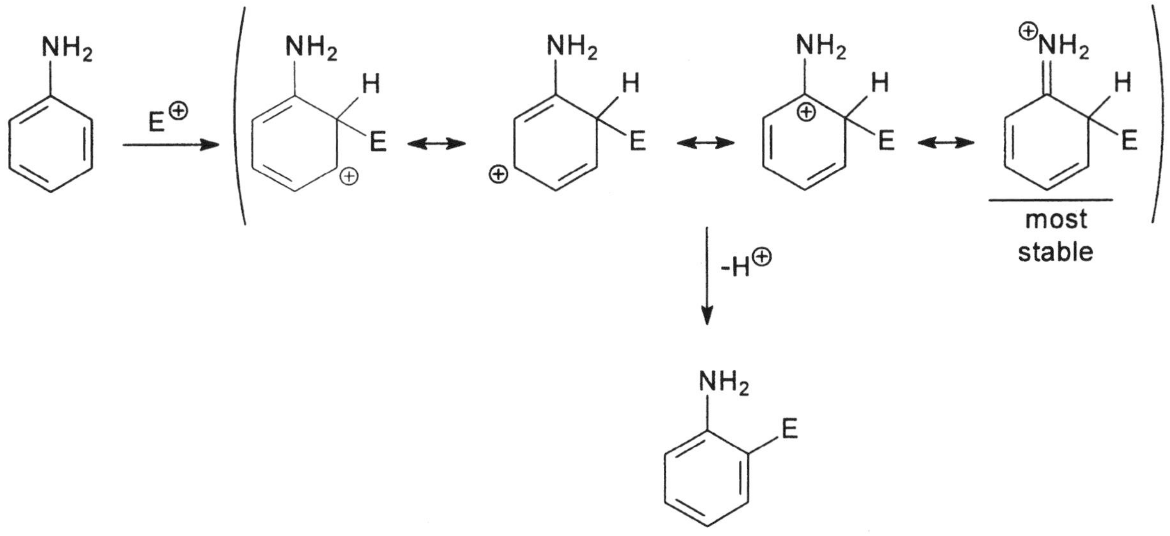

In similar fashion, *any* substituent group, in which at least one non-bonding electron pair exists on an atom directly attached to the aromatic ring will be an o,p- director and activator. These include the $-\ddot{O}H$, $-\ddot{O}R$, $-\ddot{N}HCOR$, and $-\ddot{O}COR$, groups among others. Each of these predisposes the ortho and para carbon atoms to electrophile attack and, more importantly, offers special stabilization to the resulting arenium ions. Alkyl and aryl substituent groups also cause reaction rate activation and favor o,p orientation for incoming substituents since tertiary (3°) resonance forms can be written for both, and positive charge delocalization onto the second ring can occur in the latter.

Finally, halogen substituents are o,p directors since they offer non-bonded electron pair involvement to stabilize the arenium ion. The halogens tend to be rate deactivators, however, due to the highly electronegative character of these atoms, which draw electron density from the ring, thus reducing reactivity. (*Halogens* are the only substituents which are o,p directors, but rate deactivators.)

Nitrobenzene (meta Director and Deactivator)

The nitro group (NO_2) is a *deactivator and meta director*. It *withdraws* electron density from the o and p sites on the ring, decreasing the tendency of electrophiles to attack these positions, thus making the meta carbon atoms *relatively* more attractive to electrophilic attack.

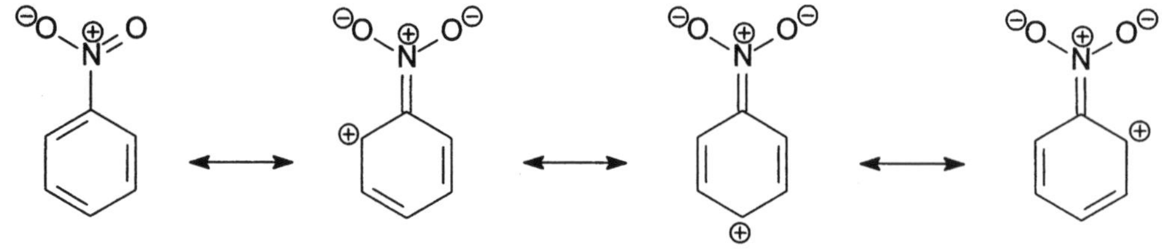

More importantly, however, if ortho or para attack were to occur, the following carbocation intermediates would be produced:

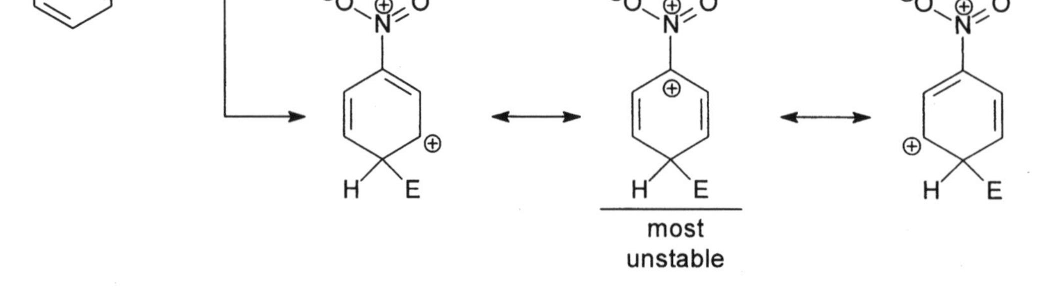

150

The underlined structures are especially *unstable* and thus would make little contribution to charge delocalization, since they place a positive charge on the carbon directly attached to the positively charged nitrogen atom. These structures are *destabilized* by like-charge repulsion on adjacent atoms. No such unstable structures result from attack at the meta carbons, so meta attack is *relatively* favored in nitrobenzene and similar compounds.

It should be noted that all of the meta directing and deactivating substituent groups listed in Fig 29-8 have either a full positive charge or a positively polarized atom *directly attached to the aromatic ring*. In addition to the nitro group, these include ketones, —CRO; aldehydes, —CHO; amides, —CONH$_2$; esters, —CO$_2$R; and carboxylic acids, —COOH; as well as nitriles, —CN. Also among the deactivators/meta directors are the trichloromethyl —CCl$_3$, and triflouromethyl —CF$_3$ groups which also have electron deficient carbons bonded to the aromatic ring and quarternary amines, —NR$_3$ $^+$X$^-$, which have a full positive charge on the nitrogen atom directly bonded to the ring. In each of these cases, the substituent group draws electron density out of the aromatic ring and also de-stabilizes the resonance structures of the arenium ion resulting from ortho- and para- electrophile attack. The result is slowed rate of reaction at these sites and therefore relative preference for meta substitution.

SUMMARY

Aromatic compounds undergo substitution rather than addition because substitution preserves the continuous cyclic pi cloud that results in aromatic stabilization.

In electrophilic aromatic substitution, an electron-deficient species, E$^+$ (the electrophile) bonds to the electron rich aromatic ring to produce a resonance stabilized carbocation called an arenium ion. Proton loss neutralizes the arenium ion and completes the reaction.

In reactions of substituted aromatics, an electrophile will bond to the ring carbon(s) having the greatest electron density and which produces the arenium ion of greatest stability. Certain substituent groups will direct electrophiles to the ortho and para carbon atoms; these generally activate the ring toward further substitution. Other groups tend to be meta directors and deactivate the ring toward further substiution.

Chapter 30
Reactions of Enolate Carbanions

Under ordinary circumstances most alkane hydrocarbons are completely non-acidic. Virtually no ionization occurs to produce a proton and a carbanion, even in the presence of the strongest bases:

The acidity of the C—H bond, however, is increased by many orders of magnitude by the presence of a neighboring carbonyl group, where resonance interaction can delocalize the negative charge of the carbanion to the oxygen atom. This greatly increases the stability and ease of formation of the carbanion, and permits it to form in the presence of even moderately strong bases:

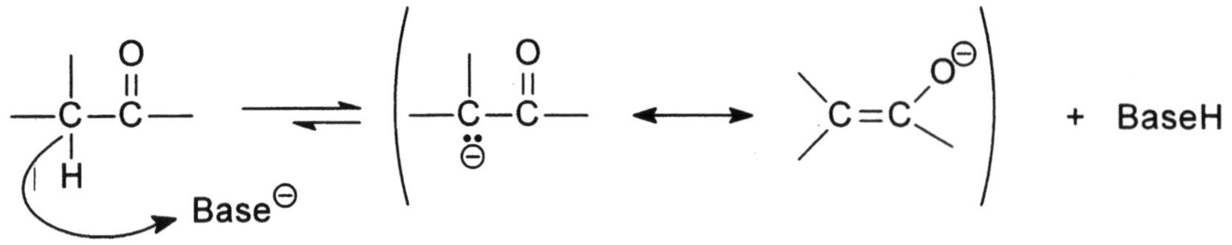

The neighboring influence of the carbonyl group thus allows the formation of an equilibrium concentration of a reactive carbanion under relatively mild conditions in the presence of common bases such as hydroxide or the oxy-anions of alcohols like sodium methoxide or ethoxide. Aldehyde, ketone, and ester functional groups stabilize and encourage the formation of carbanions in this way. The carbanions produced are called *enolates,* since the major resonance form is the anion of an alk*ene*-alcoh*ol,* also called a vinyl alcohol. Neighboring nitriles and nitro groups can also stabilize carbanions in a similar manner by charge delocalization onto the electronegative atoms (O or N) of these functional groups. For example:

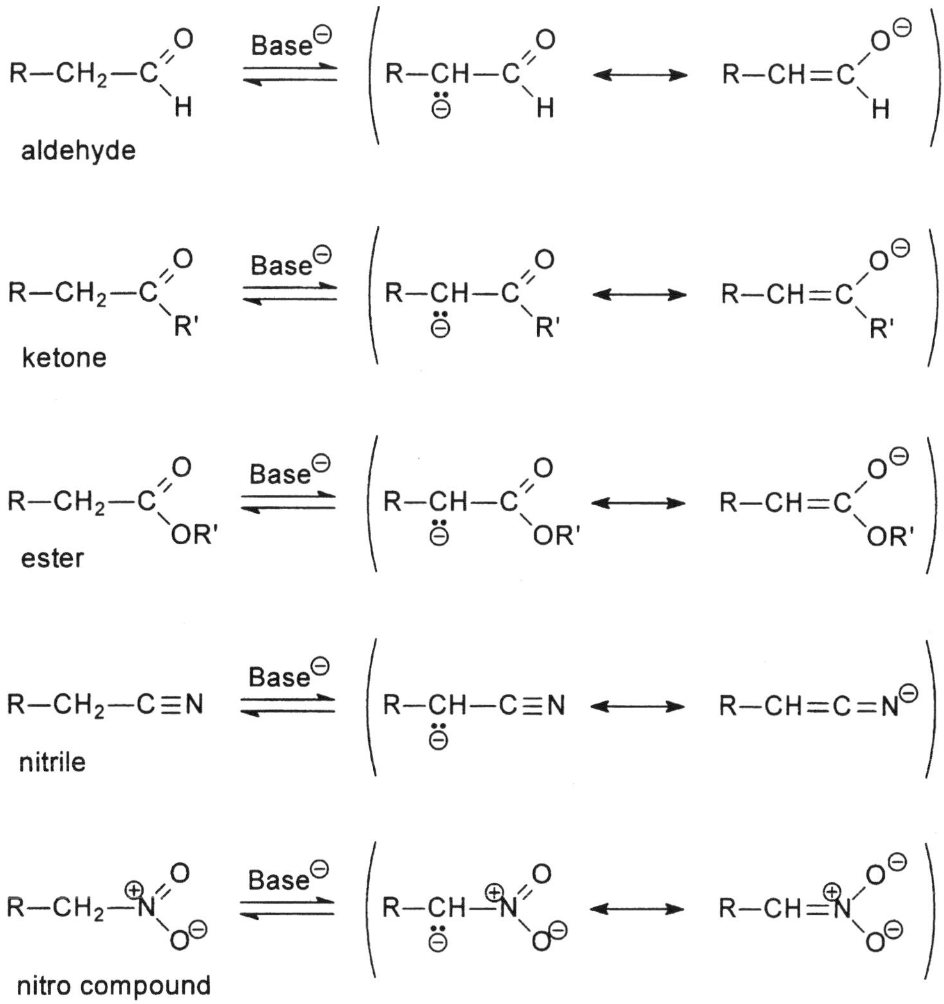

Fig 30-1: Formation of resonance stabilized carbanions.

An entire large class of chemical reactions is made possible by the easy formation of carbanions in this way. *These carbanions react in typical fashion as Lewis bases and nucleophiles, engaging in nucleophilic attack on the electron-deficient region of another molecule to form a new covalent bond.*

Aldol Reaction

An especially important example of enolate chemistry is the *aldol reaction* in which a carbanion is first formed from an aldehyde. The carbanion then engages in nucleophilic attack on the carbonyl carbon atom of another aldehyde molecule. The product of this *reductive alkylation* is called an *aldol (aldehyde-alcohol)*. The aldol often dehydrates under the reaction conditions to produce an unsaturated aldehyde as the isolated condensation product. (A *condensation* is a reaction in which a linkage forms between two molecules with expulsion of a small molecule, like water.) This and similar reactions are very important in organic synthesis because they produce new carbon-carbon bonds, increasing the size and chain length of organic compounds in a predictable and controllable manner.

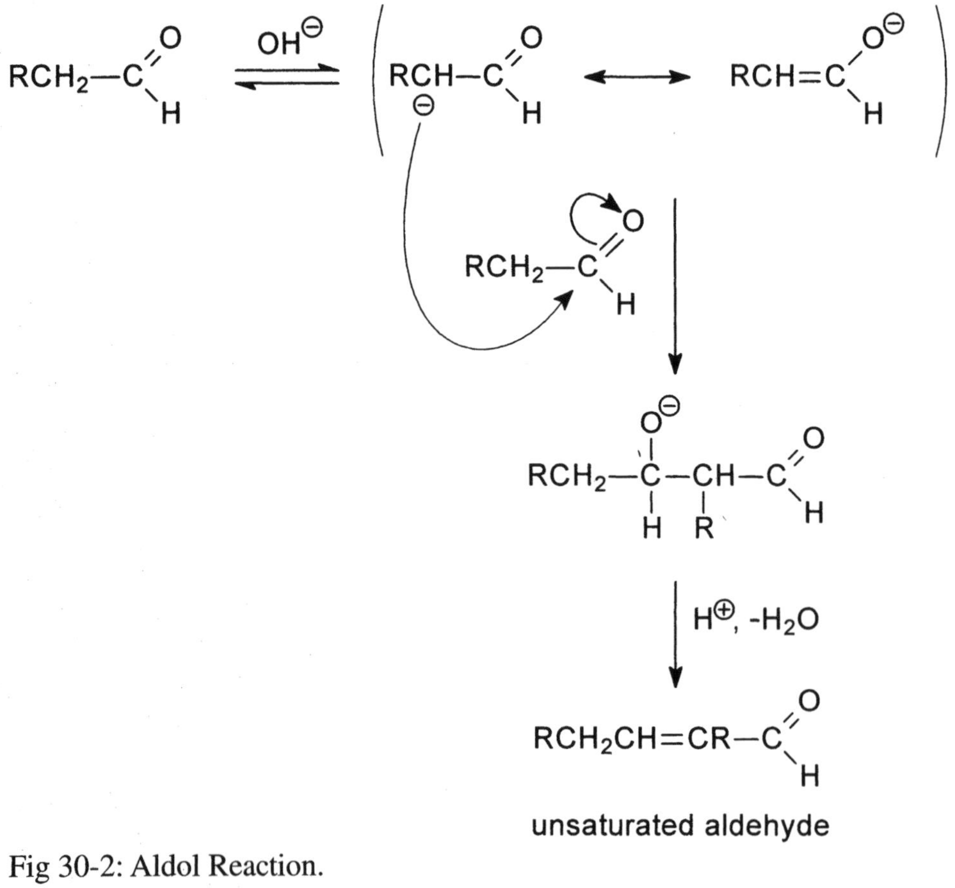

Fig 30-2: Aldol Reaction.

Crossed aldol reactions, involving two different aldehydes, can also be done, but in general this approach is synthetically useful when only *one* of the reacting aldehydes possesses acidic alpha-hydrogens.

Otherwise the product mixture becomes too complicated, with four different aldols being formed from reaction of two different carbanions with two different aldehydes. The example below shows the reaction of the carbanion derived from acetaldehyde with benzaldehyde, which has no hydrogens adjacent to the carbonyl group. A high yield of only one product is also encouraged by slowly adding *diluted* acetaldehyde dropwise to *concentrated* benzaldehyde.

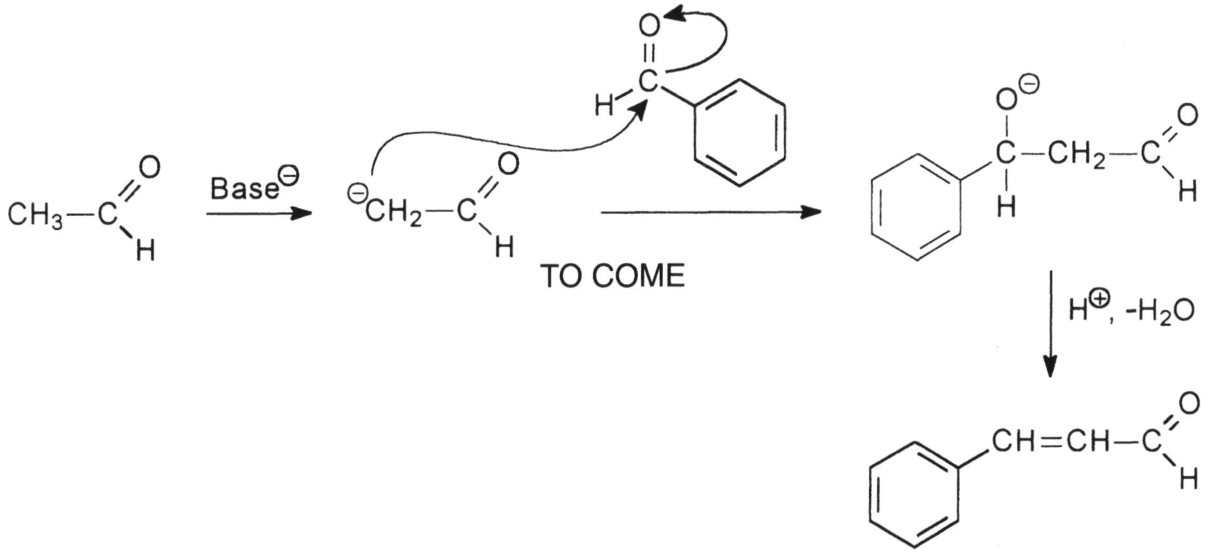

Fig 30-3: Crossed aldol reaction.

The Haloform Reaction

In another variation on this theme, the initially formed carbanion can react with halogen molecules such as Bromine, Br_2, or Iodine, I_2, to produce an α-haloketone:

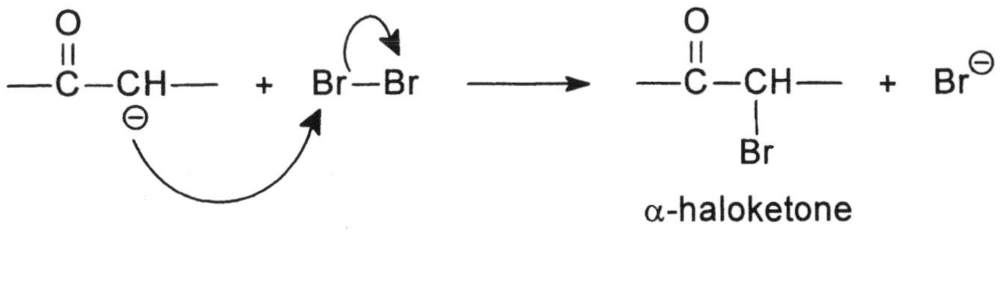

The reaction is analogous to an S_N2 substitution reaction. In the special case where a *methyl ketone* is used, this process can proceed in three successive steps introducing one, two, and then a third halogen atom to produce a trihalomethyl ketone. The trihalomethyl anion, CX_3^-, is a sufficiently good leaving group to allow formation of the carboxylic acid and the *haloform*, HCX_3, in the presence of aqueous NaOH by the usual acyl substitution mechanism (see Chapter 25). The equilibrium is pulled toward the right in this reaction because of the formation of the insoluble haloform product.

155

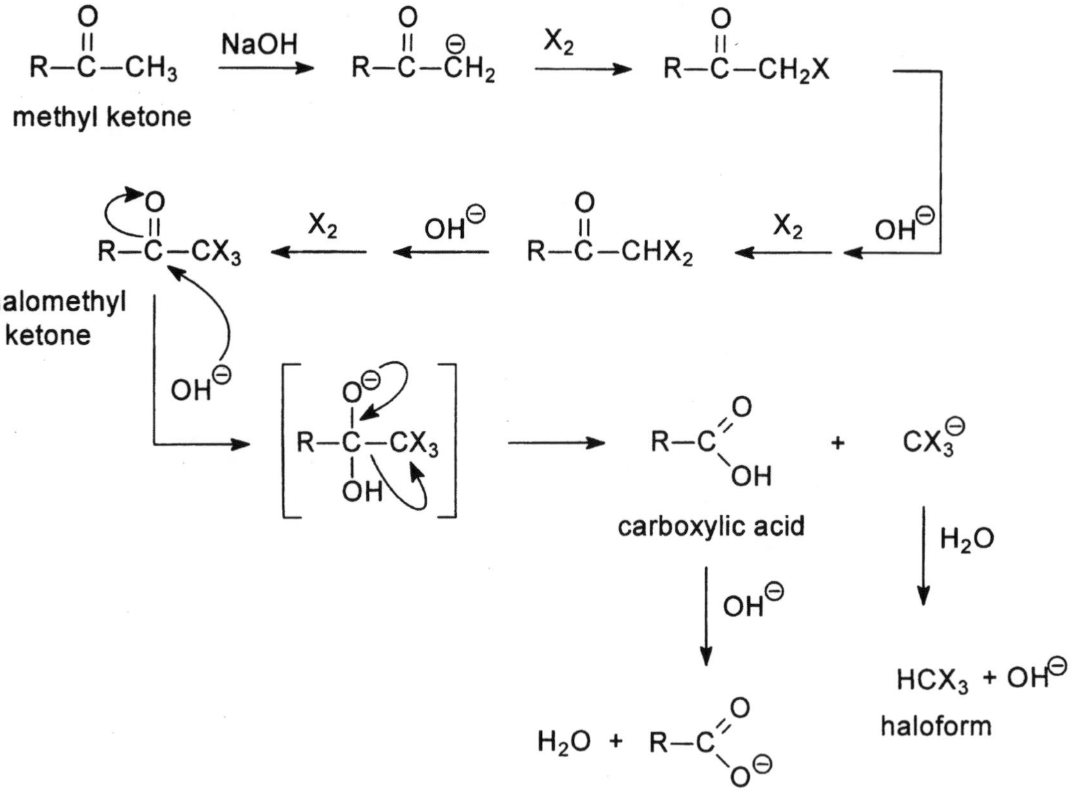

Fig 30-4: The haloform reaction.

The Iodoform Test

The haloform reaction provides a general synthetic method for the production of carboxylic acids from methyl ketones. It also provides the basis for a visual qualitative test for methyl ketones. When I_2 is used as the halogen, the products are a carboxylic acid and *iodoform,* CHI_3, which precipitates as a yellow, insoluble solid. The visual appearance of this colored precipitate provides a qualitative test for the presence of a methyl ketone, —$COCH_3$, in an organic structure. A positive *Iodoform Test* can be valuable in helping to reveal that specific structural feature in an organic compound. (Secondary alcohols linked to a methyl group will also give a positive iodoform test since the iodine reagent can oxidize the alcohol to a methyl ketone.)

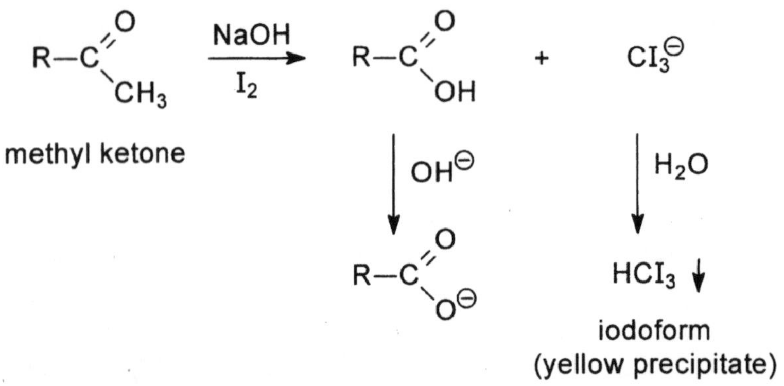

Fig 30-5: The iodoform test.

The Claisen-Schmidt Reaction

The *Claisen-Schmidt reaction* is a crossed aldol-like reaction between an aldehyde and a ketone to give a ketol, (a *keto-al*cohol) which will often dehydrate under the reaction conditions. For example:

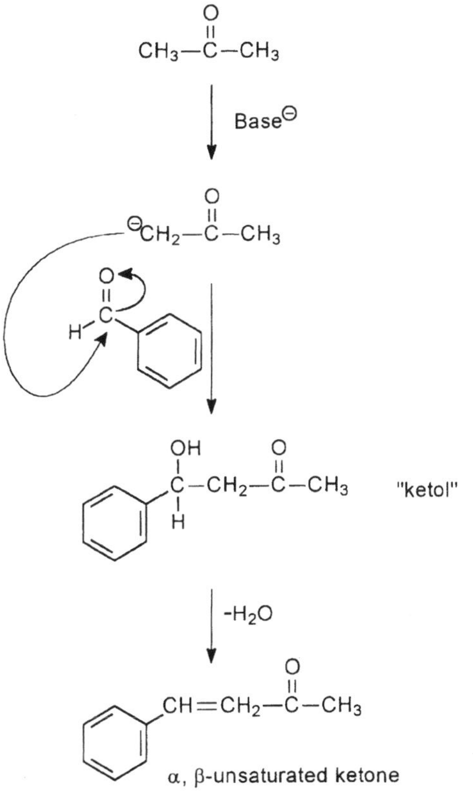

Fig 30-6: Claisen-Schmidt reaction.

Claisen Condensation

In *Claisen condensation, esters* encourage production of stabilized enolate carbanions, which form and react in the same fashion as those produced from aldehydes or ketones described above. These then attack the positively polarized carbonyl carbon of another ester molecule. In this case the product is a β-ketoester.

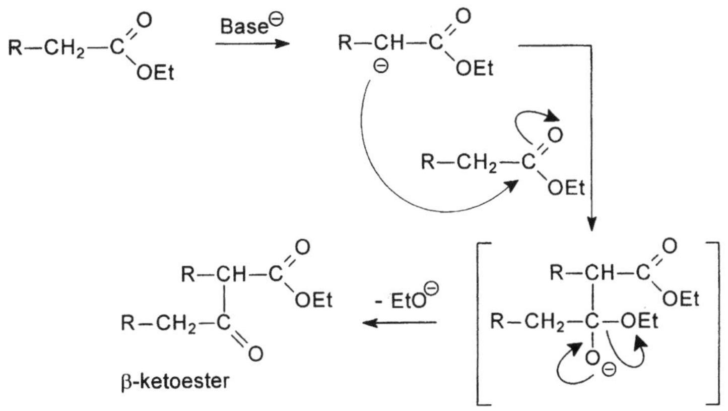

Fig 30-7: Claisen condensation.

Dieckmann Reaction

The *Dieckmann reaction* is an *intramolecular* Claisen reaction, in which the acidic hydrogen, which is the precursor to the carbanion, and the electron-deficient carbonyl carbon both occur in the same molecule. This leads, of course, to a cyclic ketoester and is very useful for the preparation of 5 and 6 membered ring structures.

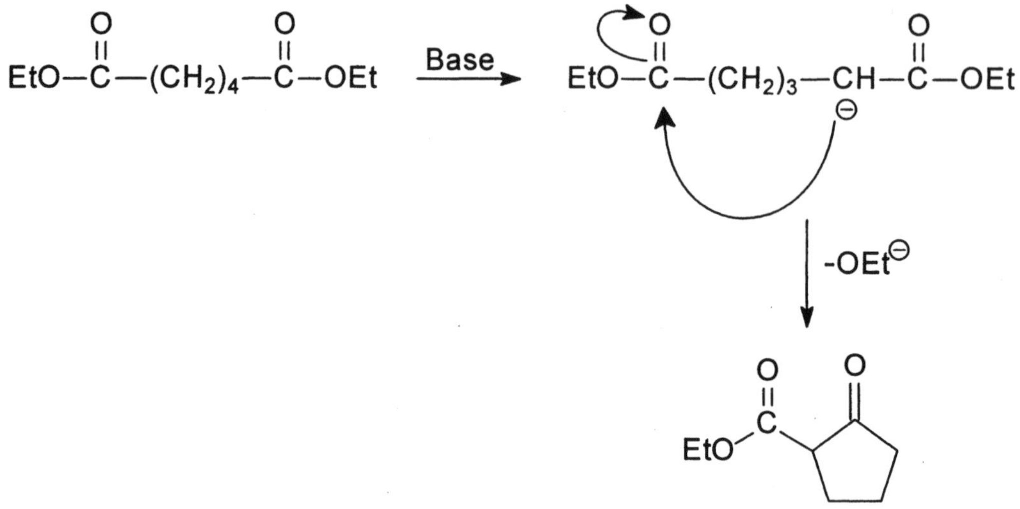

Fig 30-8: Dieckmann reaction.

Crossed Claisen Reaction

Crossed Claisen reaction between two *different* esters can also be done, preferably employing one ester which lacks acidic α-hydrogen atoms. This encourages the formation of a single product in high yield. (If two different carbanions were to form, each could attack either ester to produce a complicated reaction mixture containing as many as four different products.)

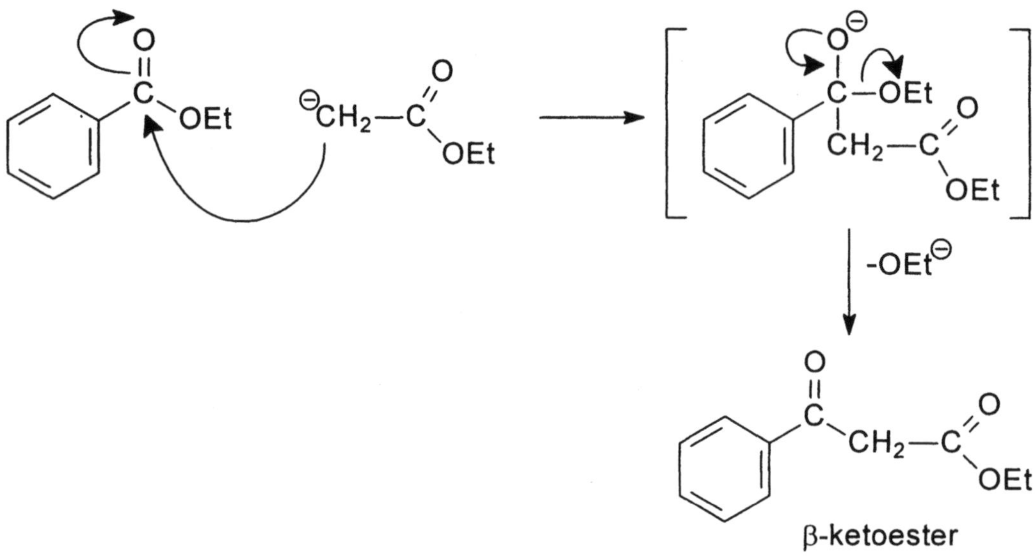

Fig 30-9: Crossed Claisen reaction.

Acetoacetic Ester Reaction

Acetoacetic ester can also be used to generate a carbanion which then participates in nucleophilic substitution at saturated carbon, for instance, with an alkyl halide. These carbanions form with special ease, since the negative charge is resonance delocalized onto *two* oxygen atoms. Alkaline hydrolysis of the ester then produces the anion of the carboxylic acid. This readily *decarboxylates* under the conditions of the reaction since the carbanion left behind after CO_2 loss is stabilized by resonance interaction with the neighboring ketone. The product of this reaction is an alkyl substituted methyl ketone. (In a variation on this theme, the carbanion can also be reacted with a carboxylic acid chloride or an α-bromo ketone to produce a β-diketone or a γ-diketone, respectively.)

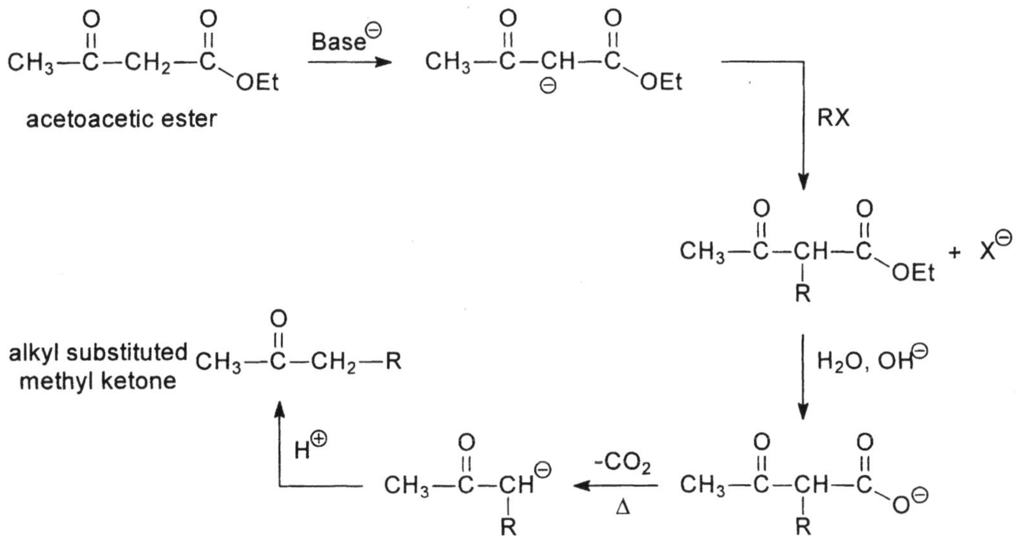

Fig 30-10: Acetoacetic ester reaction.

Knovenagle Reaction

In the *Knovenagle reaction,* acetoacetic ester is reacted with an aldehyde, like benzaldehyde, having no α-hydrogens:

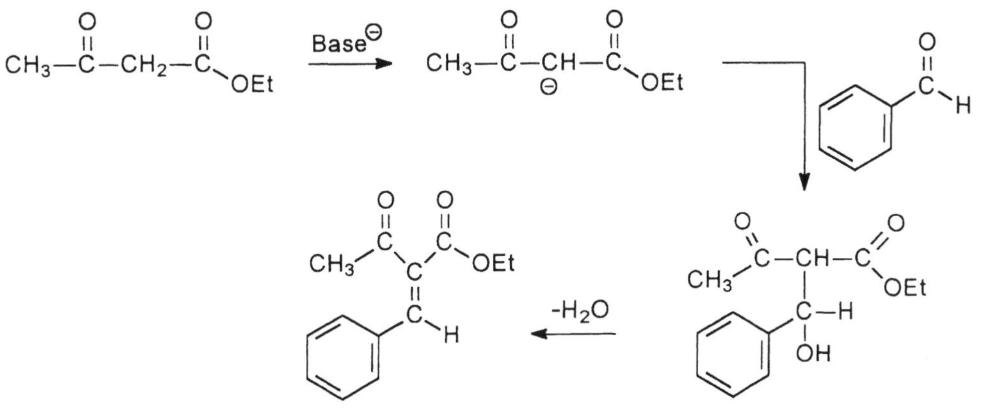

Fig 30-11: Knovenagle condensation.

Malonic Ester Reaction

Malonic ester can also form carbanions with relative ease, since the charge can be delocalized onto two oxygen atoms. This then proceeds to react with alkyl halides or tosylates, RX, by S_N2 substitution. The initially produced alkylated product can then be hydrolyzed to a dicarboxylic acid and decarboxylated to yield a desired end product.

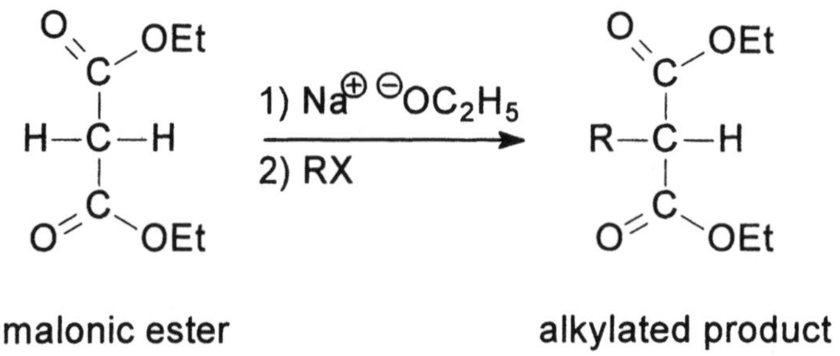

malonic ester   alkylated product

Fig 30-12: Malonic ester reaction.

Micheal Addition

Carbanions can also *add* to certain suitably substituted *alkenes*. This can occur if the structure of the alkene permits resonance delocalization of the negative charge onto a neighboring oxygen or nitrogen atom. This reaction, called *Micheal addition,* involves the addition of a carbanion, or other nucleophile, to an alkene which is directly linked to a stabilizing functional group—often an aldehyde, ketone, or nitrile. The reaction is completed by neutralization of the negative charge by a proton from the solvent. For example:

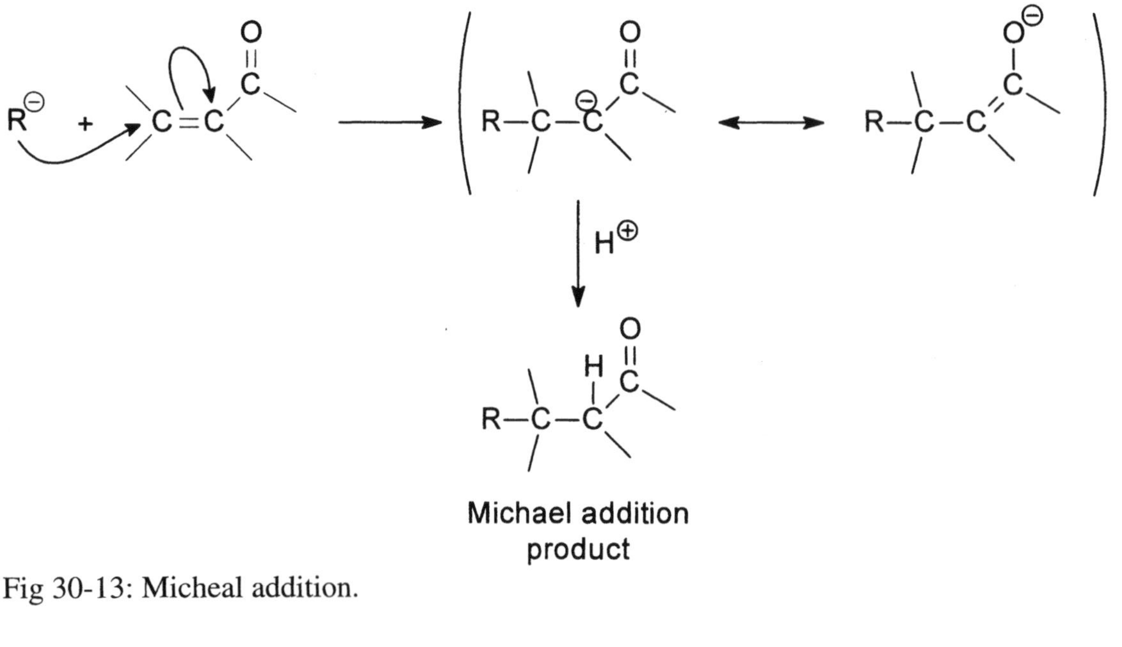

Michael addition
product

Fig 30-13: Micheal addition.

SUMMARY

All of these many reactions, and others as well, proceed by the same initial mechanistic steps. First a resonance stabilized carbanion is formed by reaction of a compound possessing an acidic α-hydrogen atom with a base. Aldehydes, ketones, esters, nitriles, and nitro compounds can do this as long as they possess at least one hydrogen atom bonded to a saturated carbon, adjacent (α) to the stabilizing functional group. This carbanion, a strong Lewis base, or nucleophile, then bonds to an electron deficient, Lewis acidic, center. These centers can include the carbonyl carbon of an aldehyde, ketone, ester, or acid chloride, a halogen molecule like Br_2 or I_2, the positively polarized carbon atom of an alkyl halide, or an appropriately substituted alkene. In many cases the conditions of the reaction result in further substitution, dehydration, or decarboxylation to produce the isolated end product.

Enolate chemistry is another example of how a wide range of organic reactions occurs via a common, simple mechanism.

It is worth noting as well that many of the reactions of enolate carbanions are similar to reactions of carbanions derived from Grignard reagents or from alkynes.

Chapter 31
The Amines

Amines contain a nitrogen atom linked by a single bond to one or more carbon atoms; they are thus organic analogs of ammonia. They are designated primary (1°), secondary (2°), tertiary (3°), and quarternary (4°) to identify the number of carbon atoms directly bonded to the nitrogen atom.

$$NH_3 \qquad RNH_2 \qquad RR'NH \qquad RR'R''N$$

ammonia   1° amine   2° amine   3° amine

Amines are named to include the word *amine* or *amino-* (prefix) along with the names of the attached organic groups. Many common historical names also occur among the amines, which include biologically and medically important compounds. Some examples of open chain, cyclic, aromatic, and complex amines are shown and named below.

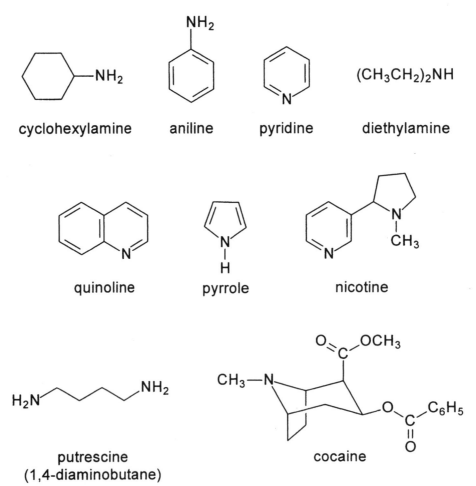

Fig 31-1: Structure of various amines.

162

The most important feature of amines is the *non-bonded electron pair* on the nitrogen atom of 1°, 2°, and 3°, amines. This electron pair is readily donated to form covalent bonds in reactions with other molecules. The amino group thus reacts as a Lewis base or nucleophile in almost all of its chemical reactions. The amines are the *bases* of organic chemistry, and are organic analogs of ammonia. Like ammonia, they react with acids in general to form salts and confer a basic pH when dissolved in water:

$$NH_3 \ + \ H_2O \ \rightleftharpoons \ NH_4^{\oplus} \ OH^{\ominus}$$

ammonium hydroxide

$$R{-}NH_2 \ + \ H_2O \ \rightleftharpoons \ R{-}NH_3^{\oplus} \ OH^{\ominus}$$

alkylammonium hydroxide

$$R{-}NH_2 \ + \ HCl \ \rightleftharpoons \ R{-}NH_3^{\oplus} \ Cl^{\ominus}$$

alkylammonium chloride

Fig 31-2: Acid-base reaction of ammonia and amines.

Properties of Amines

Shape

Alkyl amines are composed of a central nitrogen atom bonded to a total of three carbon and hydrogen atoms and possessing one non-bonded electron pair. These four attached electron pairs point to the four corners of a slightly irregular tetrahedron, defining a trigonal pyramid shape for the amine functional group.

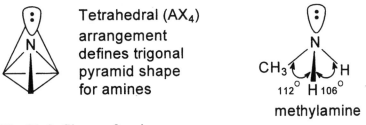

Tetrahedral (AX$_4$) arrangement defines trigonal pyramid shape for amines

methylamine

Fig 31-3: Shape of amines.

Interconversion between the two mirror images of amines is rapid at room temperature, making it impossible to isolate enantiomers (except when the nitrogen atom is held rigidly in a multicyclic ring system). Quarternary amines, in which the nitrogen is directly bonded to four *carbon* atoms, do not interconvert. They exist as stable mirror image forms which can be resolved as pure enantiomers if the nitrogen atom is bonded to four different groups:

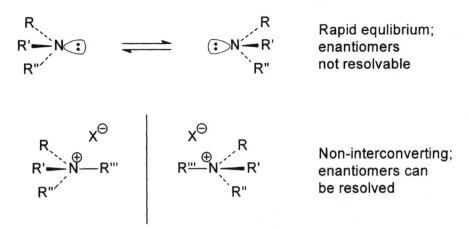

Rapid equlibrium; enantiomers not resolvable

Non-interconverting; enantiomers can be resolved

Fig 31-4: Enantiomers of quarternary amines.

Solubility and Boiling Points

The following abbreviated table shows the boiling points and water solubilities of a series of amines. Boiling points are relatively high, since 1° and 2° amines are capable of hydrogen bonding to some degree. Boiling points increase in a regular manner with the size of the molecule, and water solubilities decrease with increasing carbon content, as in all homologous series of organic compounds. (Quarternary amines are, of course, ionic compounds, and accordingly have very high water solubility and high melting points.)

1° amine	b.p. (°C)	H$_2$O Solubility
ethylamine	17	yes
n-propylamine	48	yes
n-butylamine	77	yes
cyclohexylamine	134	slightly
2° amine		
diethylamine	56	yes
di-n-propylamine	111	slightly
di-n-butylamine	159	slightly
diphenylamine	302	no
3° amine		
trimethylamine	3.5	yes
triethylamine	90	yes
tri-n-propylamine	156	slightly
triphenylamine	365	no

Fig 31-5: Boiling points and water solubility for a series of amines.

The base strength of amines is related directly to the availability of the non-bonded electron pair to function as a Lewis base. Not surprisingly, the aromatic amines like aniline are generally less basic than the aliphatic (non-aromatic) amines because of the contribution of ionic resonance forms in the former. This delocalizes the non-bonded electron pair into the aromatic ring, thus decreasing its availability to react with acidic molecules. Amines like pyrrole, in which the non-bonded electron pair is directly involved in the aromatic ring current, are especially weak bases.

cyclohexylamine

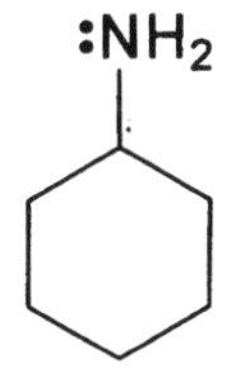

non-bonded electron pair fully localized on N atom
(no delocalization of non-bonded pair)

stronger base

aniline

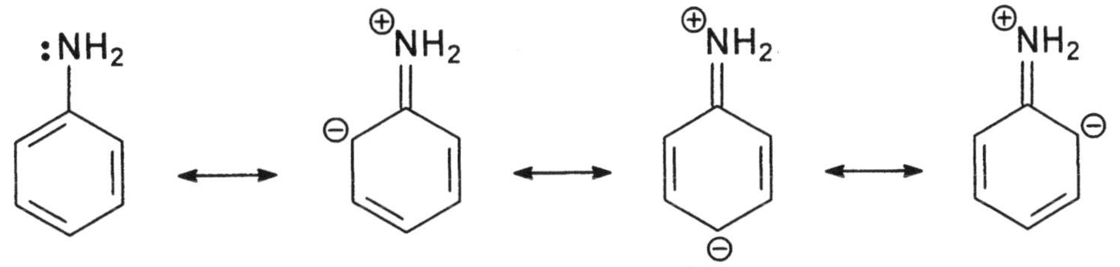

weaker base   minor contribution of ionic resonance structures
(some delocalization of non-bonded pair)

pyrrole

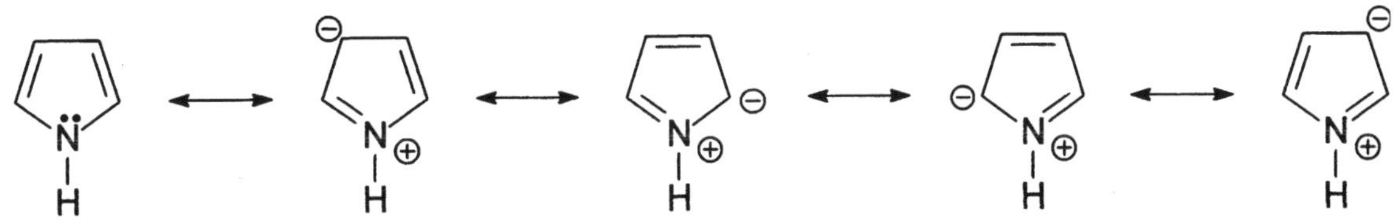

weakest base   major contribution of ionic resonance structures
(much delocalization of non-bonded pair)

Fig 31-6: Base strengths of alkyl and aromatic amines.

Preparation of Amines

Amines can be prepared by S_N2 reaction of alkyl halides with ammonia or with other amines to produce primary, 1°; secondary, 2°; tertiary, 3°; and quarternary, 4°, amines as shown below.

$$NH_3 \ + \ RX \ \longrightarrow \ NH_2R \ + \ HX$$
$$1° \text{ amine}$$

$$NH_2R \ + \ RX \ \longrightarrow \ NHR_2 \ + \ HX$$
$$2° \text{ amine}$$

$$NHR_2 \ + \ RX \ \longrightarrow \ NR_3 \ + \ HX$$
$$3° \text{ amine}$$

$$NR_3 \ + \ RX \ \longrightarrow \ NR_4^{\oplus} \ X^{\ominus}$$
$$4° \text{ amine}$$

Fig 31-7: Preparation of amines by S_N2 reactions of alkyl halides.

The yield of a desired monosubstituted (1° amine) product in this reaction can be low, however, due to the continued reactivity of the initially formed product leading to multiple substitution. Complex reaction mixtures composed of variously alkylated products can result. Several synthetic methods have been developed to circumvent this difficulty and cleanly prepare 1° amines in high yield. In one of these methods, the alkyl halide halogen is first substituted for in an S_N2 reaction by the azide ion, N_3^-. The product alkyl azide is then reduced using lithium aluminum hydride or sodium in alcohol. *This approach yields only the mono-alkylated, 1° amine, since only one carbon-nitrogen bond is present in the alkyl azide.*

$$R-X \ + \ N_3^{\ominus} \ \longrightarrow \ R-N_3 \ + \ X^{\ominus} \ \xrightarrow[\text{or} \atop \text{Na / EtOH}]{LiAlH_4} \ R-NH_2 \ + \ N_2$$

alkyl   azide   alkyl   1° amine
halide anion   azide

Fig 31-8: Production of 1° amines by reduction of an alkyl azide.

The *Gabriel synthesis* is another clever approach which cleanly produces 1° amines from the corresponding alkyl halide. In this scheme, the nitrogen anion derived from phthalimide is substituted for the halide of an alkyl halide in an S_N2 reaction. This is then reacted with hydrazine, NH_2NH_2, and the desired 1° amine is then liberated by intramolecular substitution by the neighboring acyl hydrazide accompanied by proton transfer. In this scheme as well, only the 1° (monoalkyl) amine is produced since only one carbon-nitrogen bond is present in the alkyl phthalimide.

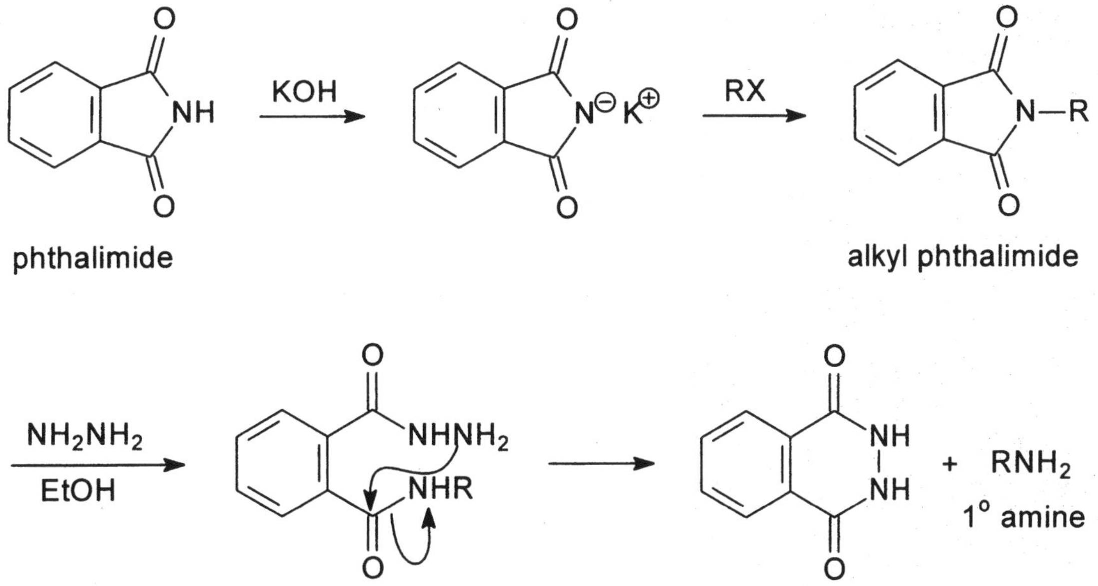

phthalimide   alkyl phthalimide

1° amine

Fig 31-9: Gabriel synthesis of primary amines.

Aromatic amines can also be produced by reduction of the corresponding nitro compound using a variety of reducing techniques. This is a common route for the production of aniline derivatives. For example:

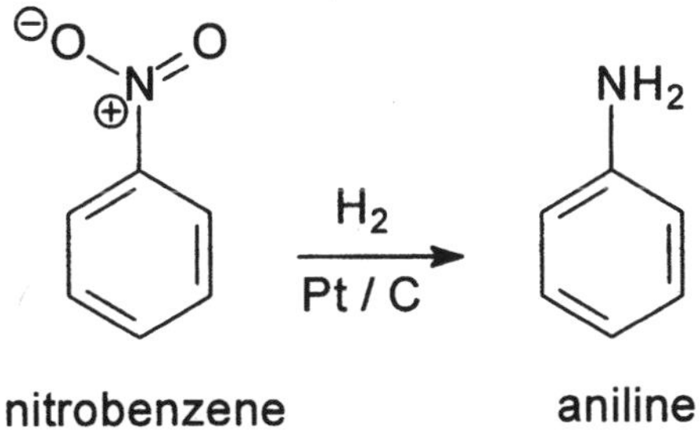

nitrobenzene   aniline

Fig 31-10: Catalytic hydrogenation
of an aromatic nitro compound.

Reductive amination can be used to convert ammonia or a 1° or 2° amine into a more highly substituted amine. The reaction sequence involves nucleophilic attack on a carbonyl compound, followed by dehydration, and finally reduction by catalytic hydrogenation or a hydride reducing agent:

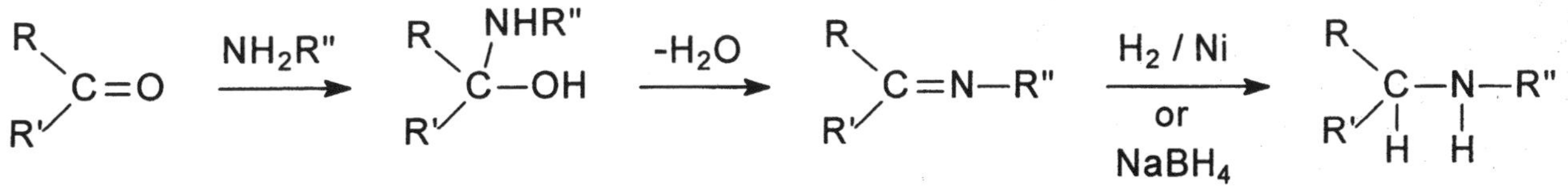

Fig 31-11: Reductive amination to produce 2° amines.

167

Nitriles can also be reduced to amines by addition of two moles of H_2 across the two pi bonds:

$$R-C\equiv N \xrightarrow[\text{catalyst}]{\text{H}_2 \atop \text{Ni}} R-CH_2-NH_2$$

And oximes, derived from aldehydes by reaction with hydroxylamine, can be reduced and dehydrated to amines, probably by the mechanistic sequence shown below:

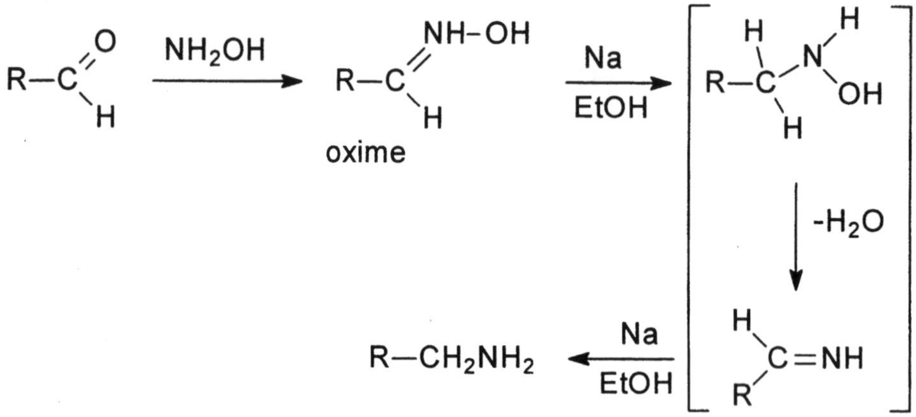

Fig 31-12: Amine production by reduction of an oxime.

(These various reduction reactions can generally be visualized as a series of steps involving uptake of hydrogen as, H·, H^-, or e^-/H^+, followed in some cases by dehydration.)

Finally, the Hoffman and Curtius rearrangements are methods for the conversion of amides and carboxylic acids into amines:

Hoffman Rearrangement

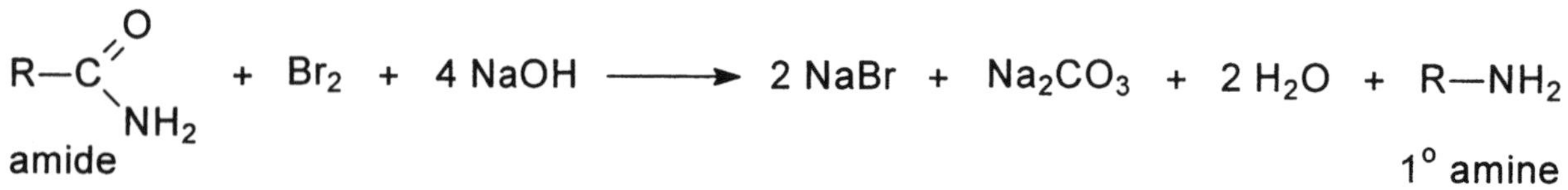

Curtius Rearrangement

168

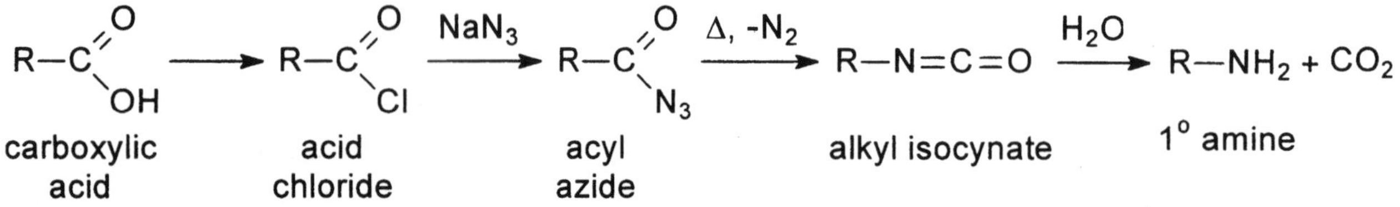

Reactions of Amines

Amines react as Lewis bases in most of their important reactions. These are primarily nucleophilic substitutions occurring at saturated carbon or acyl (carbonyl) carbon atoms. Amines also react with sulfonyl chlorides to produce sulfonamides. *The key to all of these reactions is, of course, the non-bonded electron pair on the nitrogen atom, which is donated in covalent bond formation.*

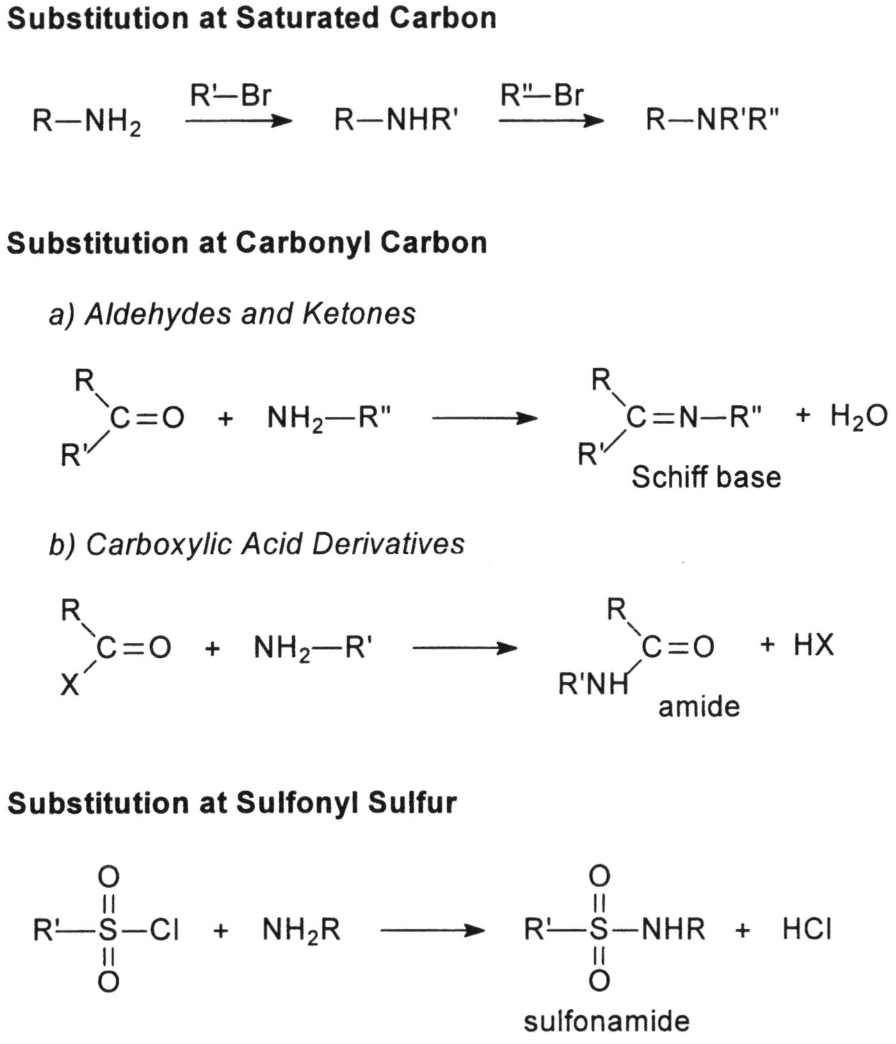

Substitution at Saturated Carbon

$$R-NH_2 \xrightarrow{R'-Br} R-NHR' \xrightarrow{R''-Br} R-NR'R''$$

Substitution at Carbonyl Carbon

a) Aldehydes and Ketones

$$\begin{matrix} R \\ R' \end{matrix}\!\!C=O \ + \ NH_2-R'' \longrightarrow \begin{matrix} R \\ R' \end{matrix}\!\!C=N-R'' \ + \ H_2O$$

Schiff base

b) Carboxylic Acid Derivatives

$$\begin{matrix} R \\ X \end{matrix}\!\!C=O \ + \ NH_2-R' \longrightarrow \begin{matrix} R \\ R'NH \end{matrix}\!\!C=O \ + \ HX$$

amide

Substitution at Sulfonyl Sulfur

$$R'-\overset{\overset{O}{\|}}{\underset{\underset{O}{\|}}{S}}-Cl \ + \ NH_2R \longrightarrow R'-\overset{\overset{O}{\|}}{\underset{\underset{O}{\|}}{S}}-NHR \ + \ HCl$$

sulfonamide

Fig 31-13: Substitution reactions of amines.

Aromatic amines can be oxidized to *diazonium salts* by nitrous acid, most commonly produced *in situ* by reaction of HCl with sodium nitrite. These diazonium salts are valuable synthetic intermediates, which can react further to produce aryl halides, nitriles, and phenols by the *Sandmeyer reaction*. Diazonium salts are also employed to produce *azo dyes* by a variant of electrophilic aromatic substitution in a process of great commercial importance:

$$ArNH_2 \xrightarrow[\text{NaNO}_2]{\text{HCl}} ArN_2^{\oplus}\ Cl^{\ominus}$$

aromatic   diazonium
amine   salt

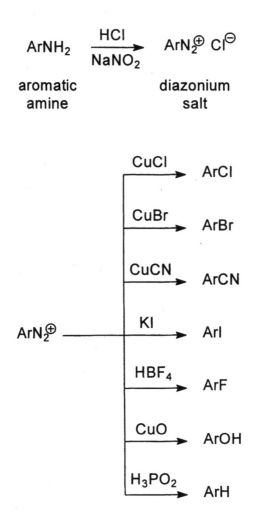

Sandmeyer Reaction

$$ArN_2^{\oplus} + Ar'X \longrightarrow Ar{-}\overset{\displaystyle N}{\underset{N}{\parallel}}{-}Ar'X$$

X = NH₂, OH   Azo dye

Fig 31-14: Production and reactions of diazonium salts.

Finally, the *Hinsberg Test* can be used to differentiate 1° , 2°, and 3° amines by a simple visual test using the reagent p- toluene sulfonyl chloride, often abbreviated tosyl chloride. (This reagent was introduced in chapter 22 for enhancing the rate of substitution reactions in alcohols.) In the Hinsberg Test, a 1° amine reacts to produce an insoluble product which becomes soluble in strong base, due to the acidity of the nitrogen-linked hydrogen. The insoluble product of a 2° amine will not dissolve in base since it has no acidic sulfonamide hydrogen. A 3° amine will react directly to produce a soluble (ionic) product. (The

Hinsberg Test is an example of an organic qualitative test used to establish the presence and nature of a functional group in a molecular structure prior to the advent of modern instrumental methods.)

THE HINSBERG TEST

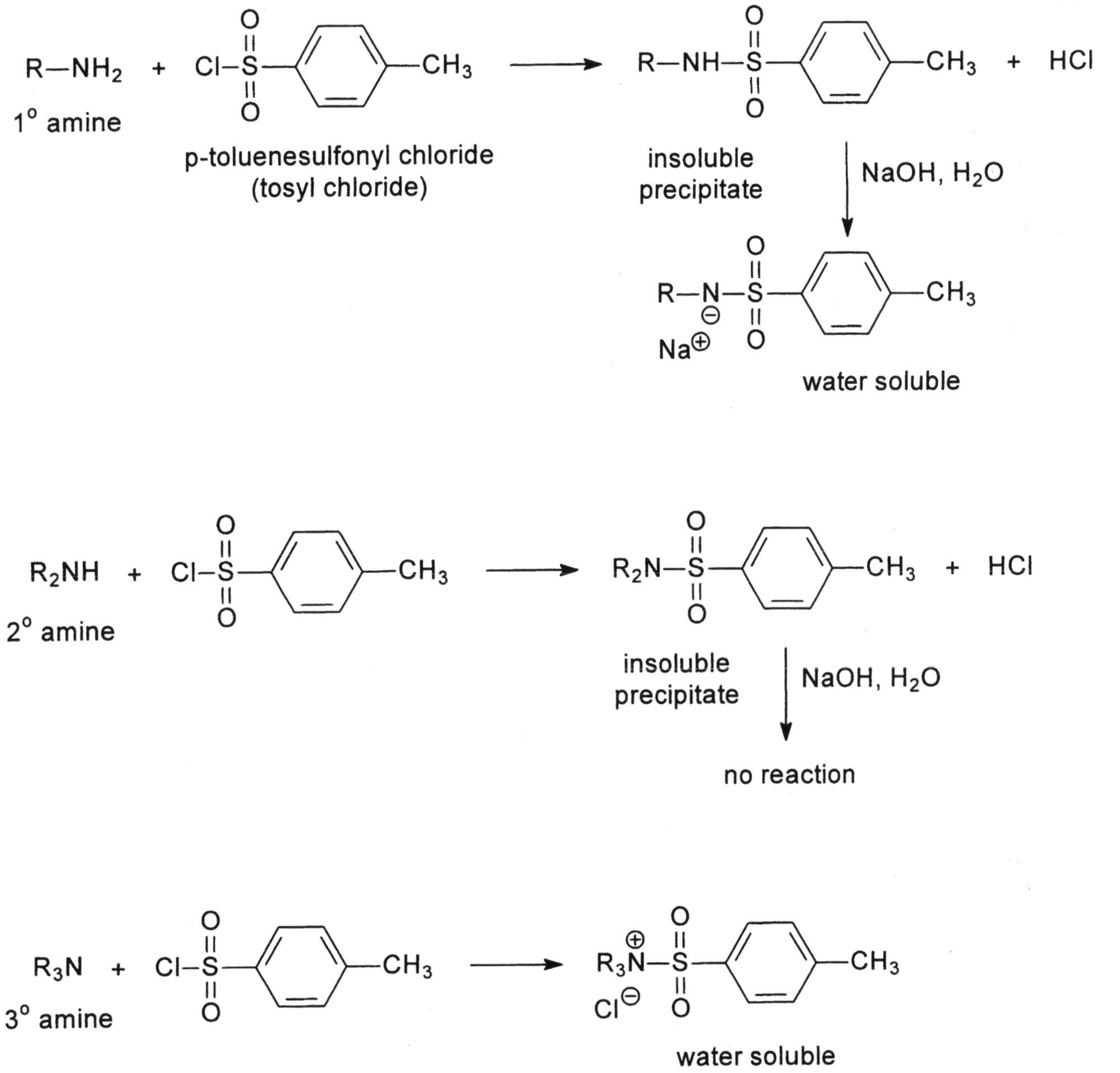

Fig 31-15: The Hinsberg Test is used to establish the presence of a 1°, 2°, or 3° amine.

Chapter 32
Overview of Reaction Mechanisms: The Production and Fate of Reactive Intermediates

The important reactive intermediates—*carbocations (carbonium ions), carbanions,* and *free radicals*—are produced in only a few ways, and once formed, proceed to react in only a few ways. This fact is one of the great unifying and simplifying features of organic chemistry. The overview which follows describes these various routes of production and reaction.

Carbocations (Carbonium Ions)

Carbocations are initially produced in only two important ways:

1. Heterolysis of a carbon-X bond where X is a halogen, protonated hydroxy, or other good leaving group. For example:

$$H_3C-\underset{\underset{CH_3}{|}}{\overset{\overset{CH_3}{|}}{C}}-I \longrightarrow H_3C-\overset{\overset{CH_3}{}}{\underset{\underset{CH_3}{}}{C}}\oplus \quad + \quad I^{\ominus}$$

$$H_3C-\underset{\underset{CH_3}{|}}{\overset{\overset{CH_3}{|}}{C}}-OH \xrightarrow{H^{\oplus}} H_3C-\underset{\underset{CH_3}{|}}{\overset{\overset{CH_3}{|}}{C}}-\overset{\oplus}{O}H_2 \xrightarrow{-H_2O} H_3C-\overset{\overset{CH_3}{}}{\underset{\underset{CH_3}{}}{C}}\oplus$$

2. Reaction of an unsaturated organic compound with a proton or other strong Lewis acid. For example:

$$\underset{H_3C}{\overset{H_3C}{>}}C=C\underset{CH_3}{\overset{CH_3}{<}} \xrightarrow{HBr} \underset{H_3C}{\overset{H_3C}{>}}\overset{\oplus}{C}-\underset{\underset{CH_3}{|}}{\overset{\overset{H}{|}}{C}}-CH_3$$

$$\underset{H_3C}{\overset{H_3C}{>}}C=O \xrightarrow{HBr} \underset{H_3C}{\overset{H_3C}{>}}\underset{\oplus}{C}-OH$$

172

Once formed, carbocations do only four important things:

1. They rapidly *rearrange* by a *hydride* or *alkide* shift from a neighboring atom *if* that will produce a more stable carbocation. For example:

2° carbocation   3° carbocation

2. They bond to an unsaturated system like an alkene or aromatic ring to form a new carbocation. For example:

3. They react with nucleophiles to form stable products. For example:

$$R^{\oplus} + X^{\ominus} \longrightarrow R-X$$

$$R^{\oplus} + H_2O \longrightarrow R-\overset{\oplus}{O}H_2 \longrightarrow R-OH + H^{\oplus}$$

4. They lose protons to form stable products. For example:

The high reactivity of carbocations is due to the instability of the electron sextet. All carbocations eventually react to reacquire a stable electron octet.

Carbanions

Carbanions are formed in only two important ways:

1. Ionization of a C-H bond by reaction with a base. For example:

$$-C \equiv C-H \ + \ Na^{\oplus} NH_2^{\ominus} \ \longrightarrow \ -C \equiv C^{\ominus} Na^{\oplus} \ + \ NH_3$$

$$-\overset{|}{\underset{H}{C}}-\overset{O}{\overset{||}{C}}- \ \xrightarrow{OH^{\ominus}} \ -\overset{|}{\underset{\ominus}{C}}-\overset{O}{\overset{||}{C}}- \ + \ H_2O$$

2. A *Grignard* type redox reaction between an alkyl halide and an active metal like magnesium or lithium. For example:

$$\overset{2e^-}{\overset{\frown}{R-Cl}} \ + \ Mg \ \longrightarrow \ R^{\ominus} Mg^{+2} Cl^{\ominus}$$

$$R-Cl \ + \ 2 Li \ \longrightarrow \ R^{\ominus} Li^{\oplus} \ + \ LiCl$$

Once formed, carbanions react in only two important ways:

1. As strong bases to abstract a proton from a wide variety of acids. For example:

$$R^{\ominus} \ + \ H_2O \ \longrightarrow \ RH \ + \ OH^{\ominus}$$

$$R^{\ominus} \ + \ R'-C \equiv CH \ \longrightarrow \ RH \ + \ R'-C \equiv C^{\ominus}$$

2. Most importantly, they react as strong nucleophiles, attacking and bonding to electron deficient centers, including carbonyl carbon atoms and positively polarized saturated carbon atoms. For example:

The high reactivity of carbanions is due to the strongly basic and nucleophilic character of the non-bonded electron pair on a carbon atom. In general, carbanions react to neutralize the negative charge on the carbon atom.

Free Radicals

Neutral free radicals are initially formed in only one important way:

1. Homolytic cleavage of covalent bonds. For example:

$$R-O-O-R \xrightarrow{\Delta} R-O\cdot + \cdot O-R$$

$$Cl_2 \xrightarrow{h\nu} 2\ Cl\cdot$$

Once formed, free radicals do only three important things:

1. They abstract atoms to produce new free-radicals. For example:

$$Cl\cdot + RH \longrightarrow HCl + R\cdot$$

175

2. They add to unsaturated systems to produce new free radicals. For example:

$$Br\cdot \ + \ \diagdown C = C \diagup \ \longrightarrow \ Br - \underset{\diagup}{\overset{\diagdown}{C}} - \underset{\diagdown}{\overset{\diagup}{C}}\cdot$$

3. They eventually combine to produce stable molecules having paired electrons. For example:

$$R\cdot \ + \ \cdot R' \ \longrightarrow \ R - R'$$

$$2 \ RCH_2 - CH_2\cdot \ \longrightarrow \ RCH = CH_2 \ + \ RCH_2CH_3$$

The high reactivity of free radicals is due to the unstable valence shell electron septet. In general, free radicals react to eventually reacquire a stable octet.

One-Step (Concerted) Reactions

Many reactions do not proceed through reactive intermediates, but rather involve the *simultaneous* breakage and formation of covalent bonds. These reactions are called *one-step, concerted,* or *simultaneous* reactions. We have described a few such reactions in this book, including substitution and elimination by S_N2 and E2 mechanisms and the Diels-Alder reaction.

Part 3
THE SPECTROSCOPIC METHODS OF ANALYSIS
Chapter 33
Introduction

The structure of organic compounds is routinely inferred from spectroscopic measurements. These methods include ultraviolet-visible (UV-vis); infrared (ir); nuclear magnetic resonance (nmr) spectroscopy, and mass spectrometry. All but the last of these techniques record the absorption of electromagnetic radiation by a target compound in a way that sheds light on its chemical *structure*. Spectroscopic methods are also used for other important purposes, especially in analytical chemistry for determining *concentration*.

The Electromagnetic Spectrum

The sun, a light bulb, or any heated object emits radiation (photons) having a wide range of energies. Photons have an associated wavelength which is inversely proportional to the energy they contain, according to the equation $E = h\nu = hc/\lambda$, where E is the energy of the photon, ν is its frequency of oscillation, and λ is its wavelength. Radiation of short wavelength has high energy while long wavelength photons carry low energy, since these quantities are inversely related. In this equation, h is a constant called Plank's constant and c is the speed of light, also a constant.

The various forms of electromagnetic radiation along with associated wavelengths are shown in the diagram below:

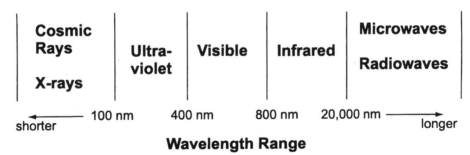

Fig 33-1: The electromagnetic spectrum.

In certain regions of the electromagnetic spectrum, especially in the ultraviolet, visible, and ir regions, photons having a narrow range of well-defined wavelengths are absorbed only by molecules which posses certain structural features, while other wavelengths are transmitted through the sample unabsorbed. Spectroscopic methods produce a record, called an *absorption spectrum,* of the wavelengths absorbed and transmitted by a given compound. A chemist can study these spectra to acquire evidence for the presence or absence of various structural features in the molecule under consideration.

Spectroscopic Methods and Molecular Structure

The various spectroscopic methods provide different types of information about molecular structure. For example, UV-visible spectra are especially useful in revealing information about the degree of unsaturation or resonance conjugation within a molecule, while ir spectra are routinely run to demonstrate the presence or absence of various functional groups, since each functional group absorbs ir radiation in a characteristic and narrowly defined frequency range. Extensive tables exist relating ir absorption frequencies to the various functional groups. Proton nuclear magnetic resonance (nmr) spectra are used to provide information about the bonding and location of hydrogen atoms in an organic structure. Carbon-13 nmr spectra reveal information about the arrangement of carbon atoms within a molecule. In all of these methods, simple, reliable, and very useful correlations exist between the absorbed radiation and specific features of molecular structure.

Chapter 34
UV-Visible Spectroscopy

In this technique, a light source is used to generate radiation over a wide range of UV and visible wavelengths. An ordinary tungsten light bulb is often used for visible light (400-800nm) and a Hg (mercury) or D_2 (deuterium) lamp is used to produce UV light having wavelengths between about 200 and 400 nm. (A nanometer, nm, is a billionth of a meter.) This radiation is then refracted or diffracted to separate it into its rainbow of component colors (wavelengths) by a prism or diffraction grating. A narrow band of this light is allowed to pass through a slit. This band changes as the prism or grating is manually or automatically turned, permitting different colors (wavelengths) to pass through the slit. This light is then focused on a solution of the target organic compound dissolved in a transparent solvent. The compound either absorbs or transmits the light, producing a signal that varies as the wavelength is systematically changed. When radiation is absorbed, the energy it contains propels an electron from a low energy (ground-state) orbital to a higher energy (excited state) orbital. The absorbed photon energy is then rapidly released, most often in the form of heat as the excited molecule returns to the ground state. Some molecules, however, lose a fraction of absorbed energy in the form of emitted light, called *fluorescence.*

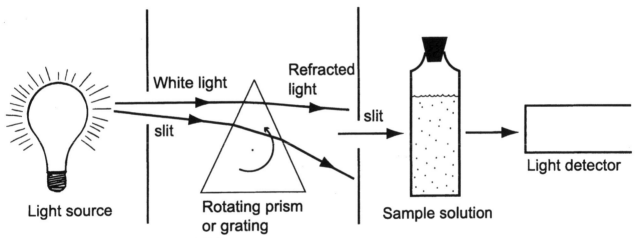

Fig 34-1: Operation of a UV-visible spectrometer.

The wavelength of maximum absorbance reveals certain properties about the absorbing molecule. As a general rule, an increase in resonance conjugation in the molecule causes an increase in wavelength (decrease in energy) of the absorbed radiation. Some examples of this rule and certain other useful generalizations of the practical use of UV-visible spectral data are summarized below.

UV-VISIBLE SPECTROSCOPY: MOLECULAR STRUCTURE AND ABSORPTION WAVELENGTH

1. Saturated compounds and simple unsaturated compounds absorb in the very short wavelength (*far UV*) region of the spectrum. Such short wavelength absorptions are generally not diagnostic for a specific type of structure and rarely offer much useful specific information about molecular

structure. Furthermore, many common solvents also absorb in the far UV, so special and elaborate techniques are often needed for measurements in this region of the spectrum. For these reasons, spectra in the far UV are not often run in routine practice. Absorption only in the far UV is a property of a wide variety of saturated and simple unconjugated, unsaturated structures.

Compound	Wavelength of Maximum Absorbance (λ max)
$CH_3CH_2CH_3$	135 nm
CH_3OH	183
$N(CH_2CH_3)_2$	227
CH_3Cl	173
$H_2C=CH_2$	171

Fig 34-2: UV absorption of saturated and simple unsaturated compounds.

2. *Ketones and aldehydes* generally show a very intense absorption in the middle to far (short wavelength) UV, and a much weaker absorption in the near (long wavelength) UV regions of the ultraviolet spectrum. The latter is due to absorption by a non-bonded electron on the carbonyl oxygen. The near-UV absorption of aldehydes and ketones can be made to disappear by the addition of a little strong acid, which protonates and ties up the non-bonded electron pair.

Compound	λ max	Extinction Coefficient, ϵ *
CH_3 \ C = O / H	180 nm 290	10,000 17
CH_3 \ C = O / CH_3	189 279	900 15

* The extinction coefficient is the proportionality constant relating light absorbance to concentration. High values of ϵ indicate strong absorption; low values indicate weak absorption.

Fig 34-3: UV absorption of simple aldehydes and ketones.

3. Simple aromatics absorb in the near (long wavelength) UV, generally between about 250 and 300 nm. The wavelength of strongest (maximum) absorption, generally called λ_{max}, increases with increasing substitution on the aromatic ring. This is especially so if the substituting groups have a non-bonded electron pair in resonance conjugation with the aromatic ring.

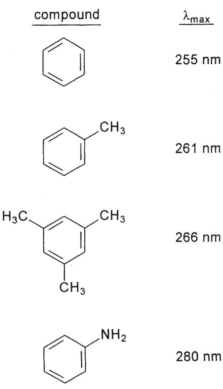

compound	λ_{max}
	255 nm
	261 nm
	266 nm
	280 nm

Fig 34-4: UV absorption of simple aromatics.

4. For a series of similar compounds, an increase in the extent of resonance conjugation results in a regular increase in the wavelength of the absorbed radiation. Increased alkyl substitution at the site of unsaturation also increases absorption wavelength, though to a smaller extent.

$$CH_3-(CH=CH)_n-C\underset{H}{\overset{O}{\diagup}}$$

compound	λ_{max}
n = 1	217 nm
2	270
3	312
4	343
5	370
6	393
7	415

Fig 34-5: Absorption of conjugated, unsaturated aldehydes.

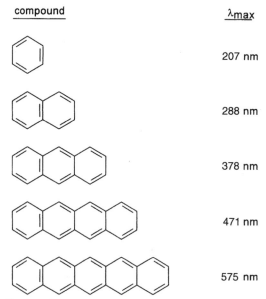

compound	λ_max
	207 nm
	288 nm
	378 nm
	471 nm
	575 nm

Fig 34-6: Absorption of polynuclear aromatic hydrocarbons.

5. Extensively conjugated molecular systems strongly absorb long wavelength *visible* radiation, even at low concentration, and are intensely colored for this reason. Virtually all dyes have extensive conjugation, generally over a large planar structure. Color in a chemical substance is due to the absorbance and removal of some fraction of visible light by the compound. The complementary color we observe is that of white light after the removal of the absorbed wavelengths. Black results if *all* visible wavelengths are absorbed. Compounds that absorb *no* visible light are, of course, white or colorless.

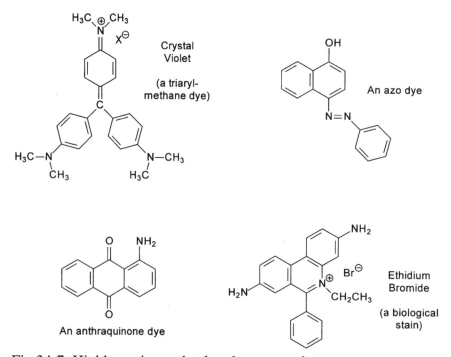

Fig 34-7: Highly conjugated colored compounds.

Chapter 35
Infra-Red Spectroscopy

Infrared spectroscopy is similar to UV-visible spectroscopy, except that the radiation is in the longer wavelength, infrared (ir) region of the electromagnetic spectrum. A broad spectrum of ir radiation is emitted by a heated cylinder of porcelain called a "glo-bar." This radiation is *refracted* (separated into component frequencies) by a prism made of a single crystal of sodium chloride. The refracted radiation is then directed onto the sample and the various frequencies are either absorbed or transmitted. The transmitted radiation is measured as the ir frequency is automatically scanned. The ir spectrum is presented most commonly as a plot of % transmittance, vs radiation frequency expressed as cm^{-1} (cycles per cm). The spectrum clearly displays which frequencies are absorbed strongly, weakly, or not at all. A typical ir spectrum is shown below. Both the prism and the cells or *cuvettes* used to contain the sample are made of salts, such as sodium chloride or potassium bromide, which do not absorb ir radiation. (Glass or similar materials would absorb all the radiation, leaving nothing left for the sample to absorb.) Solvents also have to be carefully chosen for ir transparency to minimize competing absorption. In one favored technique, the sample compound is ground up with potassium bromide, KBr. The mixture is then subjected to high pressure, which forms a transparent KBr disk or pellet containing a fine suspension of the organic compound.

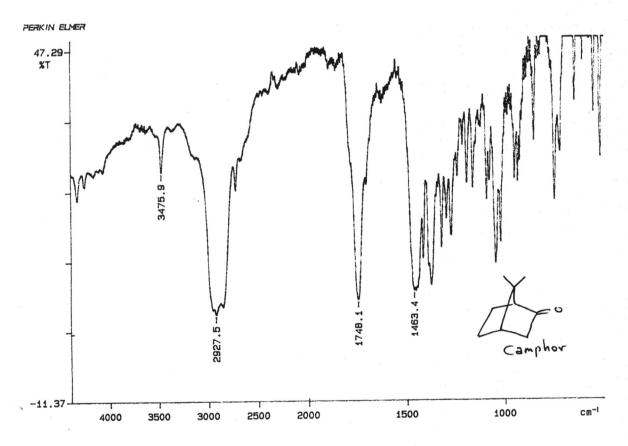

01/02/01 21:12
X: 4 scans, 4.0cm-1

Fig 35-1: A sample ir spectrum.

The energy of the absorbed photon is utilized by the molecule to increase the amplitude of oscillation (vibration) between *covalently bonded* atoms. Each type of covalent bond can only assume certain oscillation amplitudes having only certain energies. The absorption of an ir photon will occur *only* if it contains an amount of energy equal to the energy *difference* between two vibrational energy levels for that type of bond. This energetic separation is quite constant for covalent bonds of a given type. The various functional groups in organic structures are composed of these covalent bonds. As a result, the *organic functional groups characteristically absorb ir radiation over a narrow frequency range more or less independently of the structure of the rest of the molecule.* Thus, the presence of a given functional group in a molecule can be inferred by the absorption of ir radiation of a characteristic frequency. (Since in some cases several different functional groups absorb in the same range, the *absence* of a characteristic absorption is even more compelling evidence for the *absence* of that functional group.)

Extensive correlations have been tabulated relating characteristic ir absorption frequencies, usually reported in units of *wavenumbers,* with organic functional groups containing various types of covalent bonds. (Wavenumbers refer to frequency as cycles per centimeter, expressed as cm^{-1}.) Shown below is a very abbreviated version of such a table. Infrared spectroscopy is one of the most important methods available to aid in determining the structure of a newly prepared or isolated organic substance. Chemists routinely take ir spectra after synthesis or purification to establish that the desired chemical conversion of one functional group to another has indeed taken place and that the substance does not contain a contaminant containing a different functional group.

Functional Group (bond)	Frequency Range	Absorption Intensity *
Alkane (C-H)	1365 - 1395 cm^{-1}	m/s
Alkene (C = C)	1620 - 1680	v
Alkyne (C ≡ C)	2100 - 2260	v
Nitrile (C ≡ N)	2220 - 2260	v
Alcohol (O-H)	3200-3650	v/s
Carboxylic Acid (O-H) (C =O)	2500 - 3000 1710 - 1780	broad, v/s
Amide (C = O)	1680 - 1750	s
Ketone (C = O)	1690 - 1740	s
Aldehyde (C = O)	1735 - 1750	s

* M, medium; S, strong; V, variable

Fig 35-2: Characteristic ir absorption frequencies of some common functional groups.

An infrared spectrum can also be used in another valuable way. Most of the characteristic functional group absorption frequencies are found in the relatively high frequency spectral range between about 1300 and 4000 cm^{-1}. The low frequency region of the spectrum, covering roughly the range from 700 to 1300 cm^{-1}, is called the *"fingerprint"* region. Here, differences in peak position and intensity are observed *even between compounds containing the same functional groups*. The strength and frequency of absorptions seen in the fingerprint region differ depending on subtle structural details of the molecule. The fingerprint region is useful for comparing the spectrum of an unknown compound with the spectrum of one of reliably known structure. Superimposability of the two spectra is good evidence for the equivalence of these structures, often allowing a unique structure assignment to be made.

Chapter 36
Proton NMR Spectroscopy

The nuclei of hydrogen atoms (protons) in covalently bonded organic molecules behave like spinning point charges which, like all moving charges, produce a magnetic field. These protons will orient with an externally applied magnetic field generated by a large electromagnet. The spinning protons will either become *aligned* or *anti-aligned* with the applied field. These two proton populations will have an energy difference, E, proportional to the strength of the applied field. *Transitions can be induced between these states by radio frequency waves equal to the energy separation induced by the magnet.* The relationship between the frequency of the radio waves required to induce a transition between the aligned and anti-aligned states produced by a magnet of field strength H is given by the equation: $E = h\nu = 2uH$. Therefore, $\nu = 2uH/h$, where u is the magnetic moment (a constant), H is the magnetic field strength, h is Planck's Constant, and ν is the frequency of the radio wave radiation. *Thus, ν and H are directly proportional, and a change in one will produce a change in the other.*

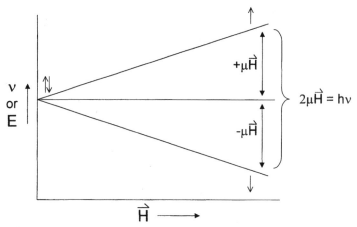

Fig 36-1: Energy separation of protons induced by an applied magnetic field.

The Chemical Shift

The protons of the various covalently bonded hydrogen atoms in an organic molecule tend to absorb energy at slightly different frequencies for a constant applied magnetic field strength. This is because the local *electron environment* near the absorbing proton creates small internal magnetic fields that add to, or subtract from, the externally applied magnetic field. Thus, the *effective* magnetic field strength felt by different protons in a molecule will vary somewhat depending on the nearby environment. This effect causes covalently bonded protons in different chemical "neighborhoods" to exist at slightly different energetic separations when aligned or anti-aligned with the applied magnetic field. They therefore absorb slightly different microwave energies to effect a transition between these states. This phenomenon is called the *chemical shift* and has the important practical consequence that *protons which are part of, or close to, the various organic functional groups absorb energy in different, predictable, and characteristic, ranges.*

186

In practice, proton absorptions are read as the parts per million (ppm) difference between the absorption of the protons of a *reference compound* and of the target protons being studied. The most commonly used reference protons are those of the compound tetramethylsilane (TMS), which is introduced into the sample being analyzed prior to the nmr spectrum being run. TMS is used because it is volatile and thus easily removed; because its protons absorb at a sharp, well-defined location; and because TMS protons absorb upfield (away from) almost all other organic protons, so that interference by the reference standard is eliminated.

Covalently bonded protons in different "environments" tend to have chemical shifts over fairly narrow, well-defined, ppm ranges. These environments relate primarily to the proximity of the absorbing proton to various functional groups within the molecule. (See Fig 36-2.) These ppm values are denoted by the Greek letter delta (δ), and vary over a range of less than 1 ppm to about 13 ppm for the covalently bonded protons found in virtually all organic compounds. The table below shows the expected chemical shift ranges for many of the hydrogen atoms commonly encountered in organic structures. Furthermore, the relative integrated areas of the absorption at each chemical shift is proportional to the relative number of protons of each type present in the compound being studied. In this way, the chemical shifts and integrated peak areas in an nmr spectrum present an easily interpreted record of the type and number of protons present in an organic structure. Since hydrogen atoms generally make an important contribution to the total structure of a molecule, a great deal of information can be derived from its nmr spectrum.

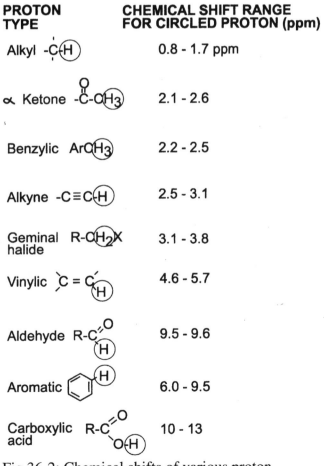

PROTON TYPE	CHEMICAL SHIFT RANGE FOR CIRCLED PROTON (ppm)
Alkyl -C(H)	0.8 - 1.7 ppm
α Ketone -C(=O)-C(H_3)	2.1 - 2.6
Benzylic ArC(H_3)	2.2 - 2.5
Alkyne -C≡C(H)	2.5 - 3.1
Geminal halide R-C(H_2)X	3.1 - 3.8
Vinylic C = C(H)	4.6 - 5.7
Aldehyde R-C(=O)(H)	9.5 - 9.6
Aromatic (H)	6.0 - 9.5
Carboxylic acid R-C(=O)O(H)	10 - 13

Fig 36-2: Chemical shifts of various proton "types" in organic compounds.

Spin-Spin Splitting

There is another important attribute of nmr spectra called *spin-spin splitting,* which further helps to diagnose molecular structure. This phenomenon is based on the fact that proton absorption is influenced by the spin of *neighboring* protons. Specifically, a given nmr absorption peak will be split into a number of sub-peaks, which depends on the number of *non-equivalent* protons directly attached to an adjacent carbon atom. This is referred to as spin-spin splitting and can help us determine the number of protons on adjacent carbon atoms in a molecule. (Neighboring protons *identically equivalent* to the absorbing protons do not cause peak splitting.)

Splitting Rules

If there is only *one* neighboring proton, it can be either aligned or anti-aligned with the applied magnetic field. The local magnetic field due to the spin of the neighboring proton will add to and subtract from the externally applied magnetic field. This *neighboring* proton alignment affects the magnetic field strength felt by the *absorbing* proton. Since the statistical likelihood of a single proton being aligned (spin up ↑) or anti-aligned (spin down ↓) is very close to 1:1, the peak at the chemical shift of the *absorbing* proton will be split into two peaks of equal intensity, called a *doublet.* The ratio of total peak areas $H_a:H_b$ will be 1:1.

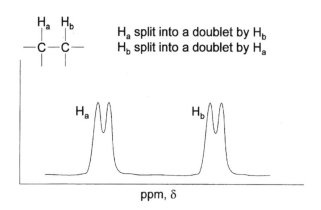

By the same token, if there are *two* protons on an adjacent carbon atom, these can both be spin up (↑↑), one up and one down (↑↓ or ↓↑), or both spins down (↓↓). The random statistical likelihood of these neighboring proton spins will be 1:2:1 respectively, and thus the absorbing peak will be split into a triplet (three sub-peaks) of relative areas, of 1:2:1. (Each of the two *one-up-and-one-down* alignments have the same influence on the absorbing proton.) The ratio of total peak areas $H_a:H_b$ will be 1:2.

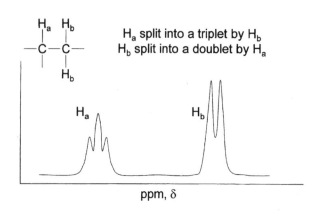

As a final example, splitting by *three* protons, which can be aligned all up (↑↑↑); one up and two down (↑↓↓ ↓↑↓ ↓↓↑); two up and one down (↑↑↓ ↑↓↑ ↓↑↑) or all down (↓↓↓), will give rise to a quartet (a peak split into four sub-peaks) in a ratio consistent with random probability, 1:3:3:1. The ratio of total peak areas $H_a:H_b$ will be 1:3.

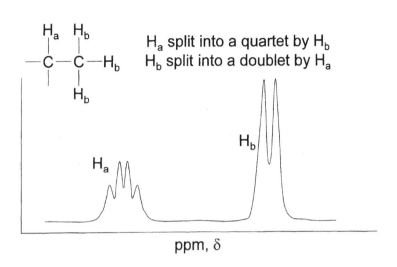

H_a split into a quartet by H_b
H_b split into a doublet by H_a

ppm, δ

In this way, the spin-spin splitting pattern in an nmr spectrum provides information about the *neighboring proton environment* of the various protons in an organic molecule. Since hydrogen atom distribution is a major feature of organic structures, the information provided by spin-spin splitting patterns provides valuable evidence for organic structure determination.

The observed spin-spin splitting patterns in nmr spectra are governed by a short list of simple rules:

1. Splitting only occurs due to the presence of *non-equivalent* protons on adjacent carbons. Environmentally equivalent protons do not split each other.
2. The separation between split peaks (called the coupling constant, J) induced by neighboring protons on each other is the same.
3. The number of split peaks is in general equal to n + 1, where n is the number of equivalent H atoms responsible for the splitting. (This simple rule holds when the splitting protons are equivalent to each other and non-equivalent to the protons being split.)
4. Splitting by two *different* sets of non-equivalent protons results in a complex splitting pattern in which each split peak caused by the first set of neighboring protons is then further split by the second set, with each obeying the n + 1 rule. (This complex pattern may be simplified in certain special cases, however.)
5. Spin-spin splitting can involve other than neighboring H atoms in aromatic systems where the circulating ring current can increase the distance over which proton spin alignment can be felt. Also, spectra taken at very low temperatures can begin to show complex splitting due to proton non-equivalence caused by restricted molecular motion in extreme cold. Indeed, nmr spectroscopy is used effectively to study such phenomena.
6. The simple rules of spin-spin splitting can break down and become much more complicated when the sets of non-equivalent protons which split each other are environmentally very similar. This occurs in cases where the difference in chemical shift between mutually splitting protons is very small and so becomes comparable to the coupling constant, J.

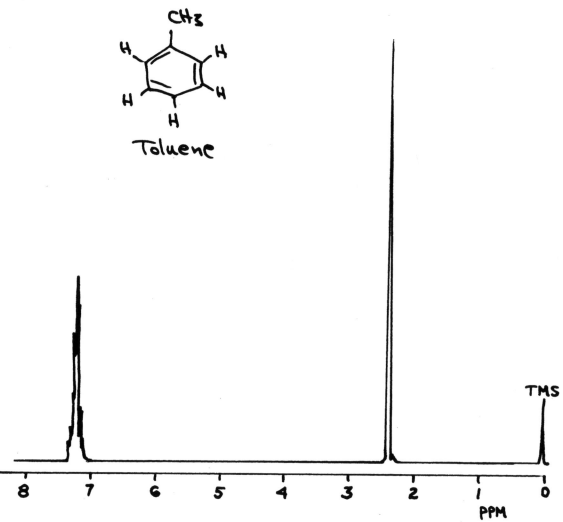

Fig 36-3: A sample proton nmr spectrum.

Chapter 37
Carbon-13 NMR

The major isotope of carbon, ^{12}C, has 6 protons and 6 neutrons, an even number. But about 1% of carbon atoms possess an extra neutron and exist as the isotope ^{13}C. This nucleus, like the proton, ^1H, has a nuclear spin, and in the presence of a strong magnetic field, spin states of two different energies are produced. Transitions between these states can be induced by irradiation with microwave photons of appropriate energy. The 1% natural abundance of ^{13}C is sufficient to express an nmr spectrum of a compound containing these isotopes.

Chemical Shifts

Chemical shifts are seen in ^{13}C nmr due to shielding or deshielding effects by atoms or atomic groupings in the environment of the absorbing ^{13}C atoms. In fact, ^{13}C chemical shifts are much *larger* than ^1H shifts seen in proton nmr, since the atom or group responsible is *closer* to the absorbing nucleus. For example, protons *adjacent* to a carbonyl group are separated from the carbonyl carbon by two bonds, while the neighboring carbon is separated by only one bond:

$$-\overset{\overset{\displaystyle O}{\|}}{C} - \overset{\overset{\displaystyle |}{}}{\underset{\underset{\displaystyle H}{|}}{C}} - \qquad -\overset{\overset{\displaystyle O}{\|}}{C} - \overset{\overset{\displaystyle |}{}}{\underset{\underset{\displaystyle |}{}}{C}} -$$

Generally, chemical shifts in ppm resulting from proximity of the various functional groups are *roughly proportional* in ^1H and ^{13}C nmr, with the magnitude of the latter being larger by a factor of roughly about 15-20.

Spin-Spin Splitting

Spin-spin splitting between adjacent ^{13}C atoms can be ignored due to the low statistical likelihood of two adjacent ^{13}C atoms (1% $\times$ 1% or about 1/10,000). ^{13}C—^1H splitting can occur but produces very complex spectra of limited usefulness. Routinely, C—H splitting is "de-coupled" (eliminated), producing a simple, easy-to-interpret, spectrum, displaying *a single sharp, un-split peak for each environmentally non-equivalent carbon atom in the molecule.*

An example of a ^{13}C spectrum is shown below:

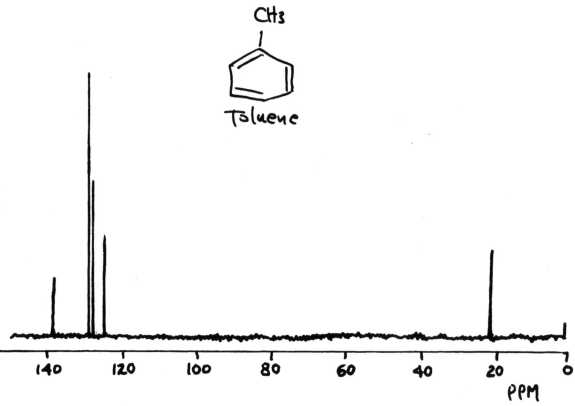

Fig 37-1: A sample ^{13}C nmr spectrum.

Comparison of ^{13}C and ^1H NMR Spectra

1. Different microwave frequencies are used to effect transitions in ^{13}C and ^1H nmr since the energy separation induced by the magnet is different for these two isotopes.
2. Peak areas (or heights) in ^{13}C spectra are not always simply proportional to the number of carbon atoms of a given chemical shift responsible for the absorption.
3. ^{13}C—^{13}C splitting can be ignored due to the low statistical likelihood of two adjacent carbon 13 isotopes.
4. ^{13}C—^1H splitting is generally "de-coupled" to produce a simple, easy- to- interpret spectrum with each *non-equivalent* carbon atom producing a sharp, un-split absorption line at a different chemical shift. Due to the very wide range of chemical shifts in ^{13}C nmr, each absorption line is generally well separated from its neighbors and is often easy to identify and assign.

Chapter 38
Mass Spectrometry

In mass spectrometry, a tiny sample of a compound is introduced into an evacuated chamber and is heated causing some of it to vaporize to a gas. These gas molecules are then hit by a stream of highly energized electrons, which are pulled across the chamber at high speed toward a positive plate (anode). Typically these electrons have a kinetic energy of 70 eV (over 1600 kcal/mole), which is very much greater than the energy needed to eject an electron from the vaporized molecules they collide with. Ejection of an electron in this manner produces a positively charged free radical called the "*parent ion*":

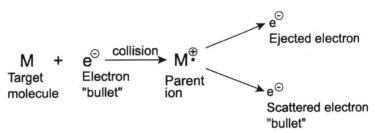

Fig 38-1: Production of the "parent ion" in a mass spectrometer.

The parent ion then breaks down further to a large extent, disintegrating into smaller positively charged fragments:

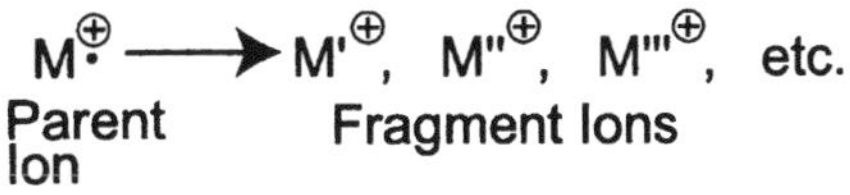

$$M^{\oplus}_{\cdot} \longrightarrow M'^{\oplus}, \ M''^{\oplus}, \ M'''^{\oplus}, \ etc.$$

Parent Ion   Fragment Ions

Fig 38-2: Fragmentation of the parent ion.

(This breakdown also produces uncharged molecules and neutral free radicals as well, but these will not interest us at this point.)

Each of these charged particles, including the parent ion, $M+$, and the various fragment ions, $M+'$, $M+''$, $M+'''$, etc., have different masses and thus a different momentum as they are compelled to move in a straight line. A perpendicular magnetic field simultaneously attempts to rotate the path of these particles. The lighter fragments (having the least momentum) are curved the most, while the heavier particles curve the least.

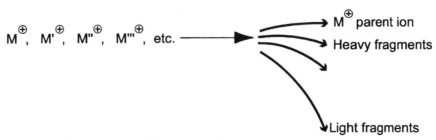

Fig 38-3: Separation of fragment ions.

The various fragment ions of different masses can be separated in this way. Each separated fragment is collected and the relative abundance of each is measured. A plot of mass/charge (m/z) vs. relative abundance is called the *mass spectrum* of the compound.

The mass spectrum can be analyzed to provide considerable information about the molecule being studied. The mass of the parent ion is the molecular weight of the compound, which is very valuable basic information. Useful evidence for molecular structure can often be obtained by noting the abundances of the various fragments, since these derive from pieces of the original molecule, which can then be reconstructed like a jigsaw puzzle. High abundance fragments often represent (relatively) more stable fragment ions. These are often produced by elimination of stable molecules, like CO_2 and H_2O, from the parent ion or from heavier fragments. Major fragments also derive from the breakdown of radical ions into neutral free radicals and simple cations. These often look like the commonly encountered reactive intermediates seen in organic reaction mechanisms. Prominent ions also derive from certain well-known cleavage reactions like retro (reverse) Diels- Alder, and ester pyrolysis reactions. Finally, the exact identity of an unknown molecule can often be firmly established by careful comparison of its mass spectrum with an entry in a collection (called a library) of standard mass spectra. Superimposibility of these spectra is compelling evidence for structural equivalence.

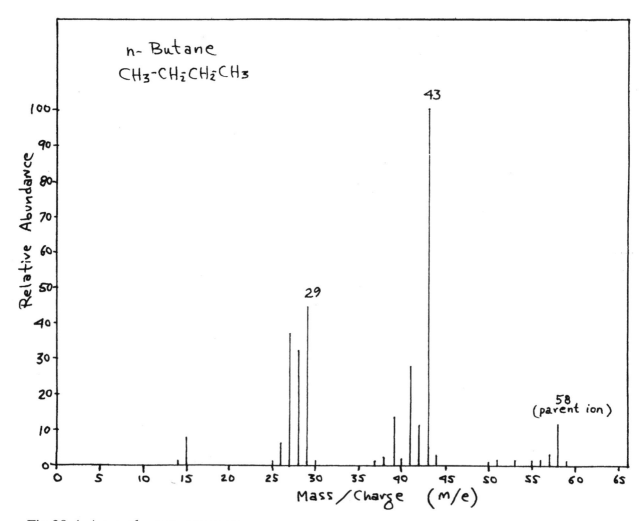

Fig 38-4: A sample mass spectrum.

194

High Resolution Mass Spectrometry and Molecular Formula

The *molecular weight* of the compound is the mass of the parent, or unfragmented, ion. The *molecular formula* can often be derived as well by employing mass spectroscopy at very high resolution, which presents the mass of the parent ion to an accuracy of four or more decimal places.

The accurate masses of atoms of the various elements are *not* exact integers. This is shown in the table below for some of the common light elements.

ISOTOPE	ACCURATE MASS
^{16}O	15.994914
^{14}N	14.003050
^{1}H	1.007825
^{12}C	12.000000

Fig 38-5: Exact mass of some common elements.

The exact molecular weight of a compound, or a poly-atomic fragment, is the sum of the exact atomic weights of the component atoms. Thus, it is possible to obtain the molecular formula of a compound by simply comparing the *exact* weight of the parent ion with entries in published standard tables relating molecular formulas to highly accurate molecular weights:

COMPOUND	EXACT MASS
C_3H_8 (propane)	44.06260
C_2H_4O (acetaldehyde)	44.02620
CO_2	43.98983
CN_2H_4	44.03746

Fig 38-6: Exact masses of some simple molecular formulas.

SUMMARY

The various instrumental methods, including UV-visible, ir, nmr, and mass spectra, can provide detailed information about molecular structure. They can reveal evidence for the degree of unsaturation; the presence or absence of various functional groups; the number and location of the hydrogen and carbon atoms in a molecule; and the molecular weight, molecular formula, and structural arrangement of the sub-units comprising an organic compound. These experimental spectra, along with tables (libraries) of standard spectra, are the most important modern tools for the determination of organic structures. An experienced chemist can often piece together these lines of evidence (along with other chemical and physical information) to establish a unique structural assignment of a complex molecule, or at least limit the candidates to a small number of possibilities.

Review Questions

This section contains questions of a general nature dealing with topics covered in the various chapters of this book. Most of these questions focus on definitions, concepts, principles, and procedures rather than with solving specific problems. Ability to correctly answer these questions will lay the foundation for solving the more focused and detailed problems generally given at the end of each chapter in reference textbooks (The number following each question refers to the chapter in this book where the answer can be found.)

PART 1 : PRINCIPLES OF ATOMIC AND MOLECULAR STRUCTURE AND REACTIVITY

1. What are the rules of writing Lewis structures, and how are they applied to describe binding in covalent compounds? (2)

2. How are organic structures represented by complete Lewis structures; by condensed structures; by bond-line representation? (3)

3. Define bond length and bond strength. Show some examples of strong and weak covalent bonds. How does bond strength relate to chemical reactivity? (5)

4. What are covalent and ionic bonds? Which representative elements form anions (negative ions) and which form cations (positive ions)? Which form covalent bonds? (2,5)

5. What is electronegativity, and how do electronegativity differences between bonded atoms affect chemical reactivity and physical properties? (6)

6. How is formal charge calculated, and what information does it provide? (7)

7. What is the Valence Shell Electron Pair Repulsion theory (VSEPR), and how is it applied to describe and predict the shape of covalent molecules? (8)

8. What are atomic orbitals and what is the sequence for filling these orbitals with electrons for the first ten representative elements (groups 1–8)? In what ways does the atomic orbital model fail to explain covalent bonding in simple organic and inorganic compounds? (9,10)

9. Explain orbital hybridization and its application to describe covalent bonding in the alkanes, alkenes, and alkynes. (10)

10. What are the common organic functional groups? Draw a general representation of each. (11)

11. What are the factors that result in attractive associations between molecules? Which are strongest? Which are weakest? What is the result of these associations on the physical properties of an organic compound, including boiling point, melting point, and solubility? (12)

12. What is the relationship between molecular mass and boiling point, melting point, and viscosity for any homologous series of organic compounds? (12)

13. Define and show examples of Lewis acids and Lewis bases. (13)

14. Describe and define constitutional isomers, stereoisomers, cis and trans isomers, diasteriomers, and enantiomers. Show examples of each. (14)

15. Define and draw the structure of a generalized meso compound. (14)

16. Describe and show an example of a concerted one-step reaction; of a two-step reaction. (16)

17. Show how carbocations, carbanions, and free radicals can be formed using arrows to describe bond cleavage. (16)

18. Explain the concept of steric hindrance and show some examples. (17)

PART 2: THE FUNCTIONAL GROUPS: PROPERTIES AND REACTIVITY

1. Define and describe alkanes. What are the characteristic physical properties and chemical reactivity of alkanes? (18)

2. Draw the various rotational states around the carbon-carbon bond of ethane. What is the relative energy and stability of each? What factors are responsible for these differences? (18)

3. What are the factors affecting the stability of the various cycloalkanes? (18)

4. What chemical reactions can be used to prepare alkanes? What reactions are undergone by alkanes? (18)

5. Define alkyl halide. What is the bond polarization in alkyl halides? How does this polarization affect chemical reactivity? (19)

6. Describe the one-step and two-step mechanisms and the products of substitution and elimination in alkyl halides. What is the effect of reactant structure, concentration, and solvent on these various reaction pathways? (19)

7. Define alkene and alkyne. In what fundamental way do these compounds differ from the alkanes? (20)

8. What are the important addition reactions of the alkenes and alkynes? (20)

9. What are the important oxidation reactions of the alkenes and alkynes? What are the products of oxidative cleavage of each class of these hydrocarbons. (20)

10. Describe the production and fate of a carbocation derived from an alkene. (20)

11. What are Makovnikov and anti-Markovnikov addition to unsymmetrical alkenes and how do these relate to carbocation stability? (20)

12. How are free radicals formed and how do they react? (21)

13. Describe all steps in the free-radical polymerization of alkenes; in the chain halogenation of alkanes; in the anti-Markovnikov addition of HBr to alkenes. What is the common characteristic of each of these reactions? (21)

14. Compare the Markovnikov (carbocation) and anti-Markovnikov (free radical) mechanism of HBr addition to unsymmetrical alkenes. Explain the difference in product isomer distribution based on the mechanism of each reaction. (21)

15. Define an alcohol and an ether. How are they similar, and in what ways do they differ in terms of physical properties? How can these differences be explained in terms of molecular interactions? (22)

16. Describe the synthesis of alcohols by hydration of alkenes and by other, more stereo-specific reactions. (22)

17. What is the mechanism of formation of a Grignard reagent? Which chemical species is oxidized? Which is reduced? (22)

18. Show the mechanism of production of various 1°, 2°, and 3°, alcohols by Grignard reaction with epoxides, aldehydes, ketones, and esters. (22)

19. Describe the nucleophilic substitution of alcohols and explain how and why reactivity is enhanced by acid catalysis and by tosylate derivatization. (22)

20. Describe the Williamson synthesis of ethers. (22)

21. Describe the preparation and reactions of epoxides. (22)

22. Write the general mechanism for the addition of a nucleophile to a carbonyl group. (23)

23. Which nucleophiles react with carbonyl compounds, and what are the various reaction products produced? (23, 25, 26)

24. Describe the reduction and oxidation of aldehydes by various reagents. (24)

25. Write the structures of the various carboxylic acid derivatives. (25)

26. Write the general mechanism of the substitution reactions of the carboxylic acid derivatives. What products result from such reactions? (25, 26)

27. What is the role of acid chlorides and other active esters? (25)

28. Describe the reaction specificity of the various hydride donating reagents. (24, 25)

29. Describe and define resonance structures. (27)

30. Define an aromatic compound. What criteria must be met if a compound is to be aromatic? (28)

31. What is the characteristic reaction of aromatic compounds? What is the mechanism of this reaction? What are a few specific examples of this reaction? (29)

32. What resonance mechanisms cause a substituent attached to an aromatic ring to be an activator/o,p-director or a deactivator/m-director? Which substituent groups fall into each catagory? (29)

33. What is an enolate ion, and how is it generally formed? (30)

34. Describe several reactions of enolate ions resulting in the formation of new carbon-carbon bonds. (30)
35. Define and show examples of 1°, 2°, 3°, and 4° amines. (31)
36. What are three reactions used to prepare amines? (31)
37. How does an amine react as a Bronsted base and as a Lewis base (nucleophile)? (31)
38. Describe the formation and reactivity of diazonium salts. (31)
39. Write equations to describe the Hinsberg Test applied to 1°, 2°, and 3° amines. (31)

PART 3: SPECTROSCOPIC METHODS OF ANALYSIS

1. How does a visible-UV spectrometer work? What information do UV-visible spectra provide relating to molecular structure? (34)
2. How does an infra-red spectrometer work? What information does it provide? (35)
3. What is the fingerprint region of an infra-red spectrum, and what information does it provide? (36)
4. What is the underlying principle behind ^{1}H and ^{13}C nmr spectra? (36,37)
5. What is *chemical shift,* and what information does it provide? (36,37)
6. What is *spin-spin splitting,* and what information does it provide? (36,37)
7. What is the principle behind mass spectrometry? What information can a mass spectrum provide? (38)

Glossary

Acetal   The functional group composed of two alkoxy groups directly bonded to a single carbon atom; a geminal diether.

Acetylene   Ethyne or an alternative name for an alkyne in general.

Activation Energy   The energy required to surmount the transition state leading to a reaction product.

Activating Group   A substituent group on an aromatic ring that causes electrophilic aromatic substitution to proceed faster than the same reaction on an unsubstituted ring.

Acylation   Introduction of an acyl group into a molecule.

Addition Reaction   A reaction in which atoms or atomic grouping add across an unsaturated covalent bond.

Alkane   A compound in which all carbon-to-carbon covalent bonds are single bonds.

Alkene   A compound having at least one carbon-to-carbon double bond.

Alkoxy   The oxygen anion of an alkyl alcohol.

Alkyne   A compound having at least one carbon-to-carbon triple bond.

Aliphatic   Any non-aromatic compound.

Alkylation   The introduction of an alkyl group into a molecule.

Alkyl Group   A group derived from an alkane by substitution of another atom or group for a hydrogen atom.

Allylic   Anything located on a saturated carbon atom directly linked to a C = C group.

Annulene   A large monocyclic hydrocarbon having alternating single and double bonds.

Aromatic   A compound having continuous resonance conjugation around a cyclic system composed of 4n + 2 alternating p orbital electrons, where n is an integer or 0.

Aryl (Ar)   An aromatic group attached to another group or atom by substitution for an aromatic hydrogen atom.

Benzyl Group   A phenyl group bonded to a saturated carbon atom.

Boat Configuration   A configuration of the cyclohexyl ring in which carbons 1 and 4 both exist above the plane defined by the other four carbon atoms.

Bond Angle   The angle defined by a central atom and two atoms directly bonded to it.

Bond Length   The distance between two covalently bonded atoms.

Bond Strength   The energy required to break a covalent bond or alternatively the energy liberated on its formation.

Bromination   A reaction in which one or more bromine atoms are introduced into a molecule.

Carbonyl Group   A carbon atom bonded to an oxygen atom by a double bond.

Chain Reaction   A reaction in which a reactive intermediate is produced along with a reaction product, resulting in many re-peating product-forming sequences.

Chair Conformation   The most stable conformation of cyclohexane, in which alternate carbon atoms occur above and below an imaginary central plane.

Chemical Shift   The variation in energy absorption by protons or carbon-13 nuclei which exist in different chemical envi-ronments in an nuclear magnetic resonance (nmr) spectrum.

Chiral   Handed; a molecule not superimposible on its mirror image and thus capable of optical isomerism.

Cis-trans Isomers   Stereo-isomers in which atoms or groups are attached to the same (*cis*) or opposite (*trans*) sides of a dou-ble bond or small ring.

Condensation Reaction   A reaction in which molecules become linked together with the expulsion of a small molecule such as water or ammonia.

Configuration   Arrangement of the atoms and groups in a stereo-isomer.

Conjugate Anion or Conjugate Base   The species produced along with the proton in the ionization of an acid: HA→ $H^+ + A^-$.

Conjugation   Extended electron distribution in a molecule resulting from overlap of p orbitals on adjacent atoms.

Glossary

Connectivity   The order in which atoms are attached to each other in a covalent molecule.

Coupling Constant   The separation of absorption due to spin-spin splitting in an nmr spectrum.

Cyanohydrin   The addition product of HCN to an aldehyde or ketone.

Deactivating Group   A group directly attached to an aromatic ring, which causes substitution reactions to proceed more slowly than on an unsubstituted ring.

Debromination   Elimination of two bromine atoms from a dibromide to produce an alkene.

Dehydration   Loss of the elements of water from a molecule.

Dehydrohalogenation   Loss of the elements of a hydrogen halide from a molecule.

Delocalization   The sharing of electron density over several atoms by resonance conjugation.

Dextrarotatory   The ability of a chiral compound to rotate the plane of polarized light to the right (clockwise).

Diastereomers   Stereo-isomers that are not mirror images of each other.

E1   Elimination: Unimolecular or First order. An elimination reaction in which the reaction rate depends on the concentration of only one reactant.

E2   Elimination: Bimolecular or Second order. An elimination reaction in which the reaction rate depends on the concentration of two reactants.

Eclipsed Conformation   A rotational conformation about a single bond in which atoms or groups on adjacent carbon atoms are directly opposed with a 0° angle between them.

Electromagnetic Spectrum   The full range of emitted photon energies with their associated wavelengths and frequencies.

Electrophile   A Lewis acid having affinity for an electron pair on another atom or group.

Enantiomer   The non-superimposible mirror image of a chiral compound.

Endothermic   A chemical reaction in which the products contain more energy than the reactants and thus draw energy from the environment as the reaction proceeds.

Epoxide   A three-membered ring composed of one oxygen and two carbon atoms.

Exothermic   A chemical reaction in which the products contain less energy than the reactants, resulting in the evolution or liberation of energy, usually in the form of heat, as the reaction proceeds.

Fatty Acid   A carboxylic acid containing a large hydrophobic (hydrocarbon) group.

Fluorination   A reaction in which one or more fluorine atoms are introduced into a compound.

Formal Charge   The charge on a covalently bonded atom. The difference between the number of valence electrons assigned to a covalently bonded atom and the number of valence electrons in a neutral atom of that element.

Functional Group   A grouping of atoms ubiquitously and commonly occurring in organic molecules. The presence, number, and proximity of these functional groups determine the chemical reactivity and physical properties of organic compounds.

Geminal (gem)   A prefix to indicate that two atoms or groups are attached to the same carbon atom.

Glycol   A compound in which two hydroxy groups are attached to adjacent carbon atoms.

Grignard Reagent   A salt containing a reactive *carbanion,* generally prepared by reacting an alkyl or aryl halide with magnesium metal.

Halogen   A group 7 element: F, Cl, Br, and I.

Halogenation   The introduction of one or more halogen atoms into a molecule.

Hemiacetal   A geminal hydroxy- ether derived from an addition reaction between an alcohol and aldehyde.

Hemiketal   A geminal hydroxy ether derived from reaction between an alcohol and ketone.

Heteroatom   Any covalently bonded atom other than carbon and hydrogen in an organic structure.

Heterocyclic   A cyclic compound containing at least one ring atom other than carbon; usually that atom is O, S, or N.

Heterolysis   Breaking a covalent bond unsymmetrically to produce charged fragments.

Homolysis   Breaking a covalent bond symmetrically to produce uncharged fragments (free radicals).

Huckle Rule   The rule stating that monocyclic planar ring compounds with $4n + 2$ p orbital electrons on adjacent atoms all around the ring will display aromatic stabilization. (In this expression n is 0 or an integer.)

Hybridization   The mixing of atomic orbitals to produce an equal number of new (hybrid) orbitals which participate in covalent bonding.

Hydration   A reaction resulting in the addition of the elements of water into a molecule.

Glossary

Hydroboration   Reaction of borane, BH_3, with an alkene to form a trialkyl borane which upon oxidation produces the anti-Markovnikov isomeric alcohol hydration product.

Hydrogen Bond   The non-covalent association of a positively polarized hydrogen atom with an electronegative atom, most often oxygen.

Hydrogenation   A reaction in which a hydrogen molecule (2 hydrogen atoms) adds to an unsaturated functional group to form a reduction product.

Hydrophilic   A polar molecule or molecular region which has an affinity for interacting with a surrounding aqueous environment. (Opposite of hydrophobic.)

Hydrophobic   A non-polar molecule or molecular region that avoids contact with a surrounding aqueous environment. (Opposite of hydrophilic.)

Infrared Region   The region of the electromagnetic spectrum including the range of wavelengths from about 2.5 to 13 microns (4000-770 cm^{-1}). The selective absorption of certain infrared wavelengths gives information about the presence or absence of specific functional groups in a molecular structure.

Intermediate (Reactive)   A transient, unstable species derived from a molecule undergoing a chemical reaction. The intermediate reacts further to produce a stable reaction product. The common intermediates include carbocations, carbanions and free radicals.

Iodination   The introduction of one or more iodine atoms into a molecule.

Ion   A chemical species bearing a net charge.

Isomers   Different molecules having the same molecular formula.

Lactam   A cyclic amide.

Lactone   A cyclic ester.

Leaving Group   The departing group or ion in a nucleophilic substitution reaction.

Levorotatory   The enantiomer of a chiral compound that rotates the plane of polarized light to the left.

Lewis Acid   A chemical species which accepts an electron pair to form a covalent bond.

Lewis Base   A chemical species which donates an electron pair to form a covalent bond.

Lipid   A biological compound characterized by hydrophobicity and solubility in non-polar media.

Markovnikov's Rule   The rule stating that the major addition product to an unsymmetrical olefin is that isomer which derives from the more stable of the possible carbocation intermediates.

Mass Spectrometry   An analytical technique in which a compound, M, is converted to its radical-cation, $M^{\dagger}$, by ejection of an electron. This radical-cation, and the various cationic fragments derived from its further breakdown, are then separated and their relative abundances counted. These data are used to provide evidence for the structure of the original molecule, M.

Monomer   Small molecules which, when covalently linked together, form a huge molecule called a *polymer.*

Newman Projection   A molecular representation in which two adjacent carbon atoms are viewed perpendicular to the plane of the paper. Newman projections are especially useful for displaying the angle formed between groups attached to the two carbon atoms.

NMR (proton)   A technique used for structure determination in which energy is absorbed by hydrogen nuclei (protons) in organic structures. The energy of these absorptions, and the relative area and details of the splitting pattern of the absorptions, can be used to reveal information about the molecular environment of the absorbing protons.

Nucleophile   An ion or atomic grouping having a pair of electrons to donate in bond formation in reaction with an electron-deficient center. The nucleophile is a Lewis base and the electron-deficient center, the Lewis acid.

Olefin   An old name for an alkene.

Optical Activity   The ability of an enantiomer to rotate the plane of polarized light.

Orbital   A region of specified geometry surrounding a nucleus in which the probability of finding an electron is high.

Oxidation   A chemical reaction involving gain of oxygen or loss of hydrogen or loss of one or more electrons.

p Orbitals   Mutually orthogonal (perpendicular) dumbbell shaped orbitals.

Paraffin   An alkane hydrocarbon.

Peroxide   A highly oxidized functional group in which two oxygen atoms are linked by a covalent bond: R—O—O—R. The weak O—O bond of peroxides provides for easy formation of free radicals.

Phenyl   Derived from benzene by substitution for a ring hydrogen atom.

Glossary

Pi Bond   The covalent bond in unsaturated compounds resulting from the overlap of p-orbitals on adjacent atoms.

Polarity   A degree of charge separation in a covalent bond due to electronegativity differences of the bonded atoms.

Polymer   A very large molecule composed of covalently bonded small molecules called *monomers*.

Primary Carbon   A carbon atom directly linked to only *one* other carbon atom.

Protonation   The bonding of a proton to an organic molecule. (De-protonation is the opposite, proton loss from an organic molecule.)

Racemic Mixture   A 50/50 mixture of the two mirror image forms (enantiomers) of a chiral compound.

Reaction Mechanism   The step-by-step sequence of bond-breaking and bond-forming events describing the progress of a chemical reaction.

Reduction   A chemical reaction involving loss of oxygen or gain of hydrogen or gain of electrons.

Resonance Structures   Individual chemical structures which, when combined as a weighted average, describe the electron distribution in a covalently bonded molecule, reactive intermediate, or complex ion.

s Orbitals   Atomic orbitals having spherical symmetry about the atomic nucleus.

Saturated Carbon   A carbon atom bonded to the maximum number (four) of other atoms.

Secondary Carbon   A carbon atom directly bonded to two other carbon atoms.

Sigma Bond   A covalent bond resulting from orbital overlap along the line joining the nuclei of the bonded atoms.

S_N1 (Substitution, Nucleophilic, First Order)   A nucleophilic substitution reaction in which the rate depends on the concentration of only *one* reacting species. Such reactions generally proceed by a carbocation mechanism.

S_N2 (Substitution, Nucleophilic, Second Order)   A Nucleophilic substitution reaction in which the rate depends on the concentration of *two* reacting species. Such reactions generally proceed by a one-step, concerted mechanism.

Solvent Effect   The influence exerted by a reaction solvent on the rate or other outcome of a chemical reaction.

Spin-Spin Splitting   The pattern of peak splitting seen in a proton nmr spectrum caused by the nuclear spin orientations of protons attached to neighboring carbon atoms.

Staggered Configuration   The configuration resulting from rotation around a carbon-carbon single bond in which all atoms or groups attached to the two carbons are separated by the maximum angle (60°).

Stereoisomers   Isomers which have atoms connected in the same sequence but differ in shape or spatial configuration.

Steric Hindrance   The phenomenon of atoms or groups in molecular species interfering with each other due to close proximity.

Substitution Reaction   A reaction in which one atom or group replaces another.

Tertiary Carbon   A carbon atom bonded to three other carbon atoms.

Torsional Strain   Instability in a molecular structure resulting from the eclipsing of atoms or groups bonded to adjacent atoms.

Transition State   The highest energy structure that reactant molecules must assume before proceeding on to reaction products.

Unsaturation   An atomic grouping in which *less than four* atoms are bonded to a carbon atom; includes all functional groups possessing multiple bonds.

Vicinal (vic)   Two atoms or groups bonded to *adjacent* atoms.

Vinyl   An atom or group directly bonded to an olefinic (alkene) carbon atom.

Visible-UV Spectrum   The spectrum describing the absorption and transmission of radiation as the wavelength is varied from about 200 to 800 nm.

Wavelength   The trough-to-trough length of photon oscillation of electromagnetic radiation.

Zeitzev Rule   The rule that the isomeric alkene most highly substituted by carbon atoms will predominate in an elimination reaction.

Zwitterion   A covalently bonded structure containing both positive and negative charges.

Index

absorption spectrum, 177
acetoacetic ester reaction, 159
acetylenes (alkynes), 28
acid ionization, 15
active esters, 123
acylation of aromatics, 145
addition of halogen acids to alkenes, 85
addition of molecular halogens to alkenes, 85
addition of water to alkenes, 85
addition reactions of aldehydes, 109
addition reactions of alkenes, 85
addition reactions of ketones, 109
addition reactions, 61
alcohols, 30, 39, 100
aldehydes, 31
aldol reaction, 154
alkali metals, 3
alkanes (paraffins), 27, 67
alkenes (olefins), 28, 84
alkide shift, 173
alkyl halide stucture, 78
alkyl halides, 30, 76
alkylation of aromatics, 144
alkynes (acetylenes), 24, 28, 84, 91
amide hydrolysis, 120
amide synthesis, 130
amines, 34, 39, 162
ammonia, 162
aniline, 149, 165
annulene, 138, 146
anthracene, 138
Anti-Markovnikov addition, 96
arenium ion, 141
aromatic compounds, 29, 135
aromatic substitution, 44
aromaticity, 135
aspirin, 116
atomic orbitals, 21
azide anion, 166
azulene, 147

barbituric acid synthesis, 121
benzoyl peroxide decomposition, 94
benzyl carbocation, 133
boiling point, 35
bond length, 11
bond polarity, 12
bond strength, 11
bond-line molecular representation, 68
boron trifluoride, 145
Bronsted acid, 41
Bronsted base, 41

carbanions, 54, 62, 174
carbene, 63
carbocation (carbonium) ions, 53, 60, 172
carbocation addition to alkenes), 87
carbon-13 NMR, 191
carbonium ions (carbocations), 60, 172
carbonyl group reactions, 128
carbonyl-containing functional groups, 31
carboxyl group, 7
carboxylic acid derivatives, 33
carboxylic acids, 33, 39, 116
catalyic hydrogenation of alkenes, 85
catalyst, 74
catalytic hydrogenation of alkenes, 88
catalytic hydrogenation of ketones, 113
chain reaction, 60
chair conformation, 73
charge delocalization, 149
charged intermediates, 60
cholesterol, 8
cis/trans configurations, 46
Claisen condensation, 157
Claisen-Schmidt reaction, 157
cocaine, 162
combustion of alkanes, 75
condensation reaction, 154
condensed molecular formula, 68
constitutional isomers, 45
covalent bonds, 4, 10
crossed Claisen reaction, 158
Curtius rearrangement, 168
cycloalkanes, 71
cyclohexane, 73
cyclohexylamine, 165
cyclopentadienyl anion, 139
cyclopropenium, 139

dextrarotation, 49
diasteromers, 49
diazonium salts, 169
Dieckmann reaction, 159
Diels-Alder reaction, 58
dissolving metal reduction of ketones, 113

E2 elimination, 58
eclipsed conformation, 64, 71
electromagnetic spectrum, 177
electron delocalization, 131, 149
electronegativity, 12
electrophilic aromatic substitution, 141
elimination reactions of alkyl halides, 82
elimination reactions, 61

Index

enantiomers, 48
enolate carbanions, 152
epoxidation of alkenes, 86
epoxide, 103, 107
ester hydrolysis, 120, 130
ethers, 31, 100, 107

ferrocene, 147
formal charge, 16
free radical reactions, 93
free radicals, 54, 59, 175
free-radical polymerization, 98
functional groups, 27
furan, 138

Gabriel synthesis of amines, 121
galvinoxyl, 133
Grignard reaction, 55, 102, 117, 174

haloform reaction, 155
halogenation of alkanes, 75, 95
halogenation of aromatics, 142
halogens, 3
halohydrin formation, 86
hemiacetal formation, 110
heteroatoms, 38
heterocycles, 146
heterolytic bond cleavage, 15
heterolytic bonding, 53
Hinsberg test, 170
Hoffman rearrangement, 168
homolytic cleavage, 54, 59
Huckle Rule, 137
hydration of alkenes, 85
hydration of ketones and aldehydes, 110
hydrazones, 111
hydride attack of ketones, 113
hydride donors, 113
hydride shift, 173
hydroboration, 102
hydrocarbons, 27
hydrogen bonding, 15

imidazole, 138
infra-red spectroscopy, 183
iodoform test, 156
isomers, 45

keto-enol tautomerization, 92
ketones, 32
Knovenagle reaction, 159

levorotation, 49
Lewis acid, 41
Lewis base, 41
Lewis molecular structure, 6
Lindlar catalyst, 92
lithium aluminum hydride, 126

magnesium halide carbanion, 102
malonic ester carbanion, 133
malonic ester reaction, 160
Markovnikov mechanisms, 96
mass spectrometry, 193
m-directors, 148
melting points, 37
mercaptans, 39
meso compounds, 50
meta- isomers, 148
Micheal addition, 160
MMPP, 89, 107
molecular formula, 68
molecular shape, 18, 65

naphthalene, 133, 147
Newman projection, 72
nicotine, 162
nitration of aromatics, 143
nitriles, 153, 168
nitrobenzene, 150, 167
non-benzenoid aromatics, 146
nucleophile, 76
nucleophilic reactions, 15
nucleophilic substitution of alcohols, 105
nucleophilic substitution of alkyl halides, 76

o, p-directors, 148
octet rule, 3
olefins (alkenes), 28
oleic acid, 116
one-step (concerted) reactions, 176
one-step mechanisms, 58
one-step SN2 mechanism, 77
optical isomers, 47
orbital hybridization, 23
ortho- isomers, 148
osmium tetroxide oxidation of alkenes, 88
oxidation of alcohols, 115
oxidation of alkenes, 86
oxidation of carbonyl compounds, 112
oxidation reactions of alkenes, 88
oxidative cleavage of alkenes, 86, 89
oximes, 111
oxiranes, 89, 107
oxymercuration/demercuration, 102
ozone oxidation of alkenes, 90

para- isomers, 148
paraffins (alkanes), 27, 67
PCC, 115
penicillin deactivation, 121
permanganate oxidation of alkenes, 88
phenanthrene, 138
phthalimide, 167
physical properties, 35
plastics, 99
polyamides, 124

Index

polymethylmethacrylate, 99
polypeptides, 125
polyvinyl chloride (PVC), 99
proteins, 124
proton NMR spectroscopy, 186
proton transfers, 44
protonation, 44
puckering, 73
putrescine, 162
PVC (polyvinyl chloride), 99
pyridine, 138, 147, 162
pyrrole, 138, 147, 162, 165

quinoline, 162

radical anion, 62
radical cation, 62
recemic mixture, 50
redox reactions, 112
reduction of carbonyl compounds, 112
reductive alkylations, 103
reductive amination, 167
resonance structures, 131

Sandmeyer reaction, 169
saturated hydrocarbons, 67
Schiff base formation, 110

SN2 substitution, 58
spectroscopic methods, 177
spin-spin splitting, 188, 191
splitting rules, 188
staggered conformation, 64, 71
stereochemistry, 45
steric hindrance, 64, 80
steroisomers, 46
substitution reactions of aldehydes, 109
substitution reactions of ketones, 109
substitution reactions, 61
substitution/addition reactions, 44
sulfonation of aromatics, 144

torsional strain, 72
tosyl chloride, 106
transesterification, 130
transition state, 57
tropylium ion, 139
two-step SN1 reaction mechanism, 77

uv-visible spectroscopy, 179

valence, 1, 3
vinyl alcohol, 91, 153
viscosity and molecular size, 36
VSEPR, 18

RAPID LEARNING AND RETENTION THROUGH THE MEDMASTER SERIES:

CLINICAL NEUROANATOMY MADE RIDICULOUSLY SIMPLE, by S. Goldberg
CLINICAL BIOCHEMISTRY MADE RIDICULOUSLY SIMPLE, by S. Goldberg
CLINICAL ANATOMY MADE RIDICULOUSLY SIMPLE, by S. Goldberg
CLINICAL PHYSIOLOGY MADE RIDICULOUSLY SIMPLE, by S. Goldberg
CLINICAL MICROBIOLOGY MADE RIDICULOUSLY SIMPLE, by M. Gladwin and B. Trattler
CLINICAL PHARMACOLOGY MADE RIDICULOUSLY SIMPLE, by J.M. Olson
OPHTHALMOLOGY MADE RIDICULOUSLY SIMPLE, by S. Goldberg
PSYCHIATRY MADE RIDICULOUSLY SIMPLE, by W.V. Good and J. Nelson
CLINICAL PSYCHOPHARMACOLOGY MADE RIDICULOUSLY SIMPLE,
        by J. Preston and J. Johnson
ACUTE RENAL INSUFFICIENCY MADE RIDICULOUSLY SIMPLE, by C. Rotellar
USMLE STEP 1 MADE RIDICULOUSLY SIMPLE, by A. Carl
USMLE STEP 2 MADE RIDICULOUSLY SIMPLE, by A. Carl
USMLE STEP 3 MADE RIDICULOUSLY SIMPLE, by A. Carl
BEHAVIORAL MEDICINE MADE RIDICULOUSLY SIMPLE, by F. Seitz and J. Carr
ACID-BASE, FLUIDS, AND ELECTROLYTES MADE RIDICULOUSLY SIMPLE, by R. Preston
THE FOUR-MINUTE NEUROLOGIC EXAM, by S. Goldberg
MEDICAL SPANISH MADE RIDICULOUSLY SIMPLE, by T. Espinoza-Abrams
THE DIFFICULT PATIENT, by E. Sohr
CLINICAL ANATOMY AND PHYSIOLOGY FOR THE ANGRY HEALTH PROFESSIONAL,
        by J.V. Stewart
CONSCIOUSNESS: HOW THE MIND ARISES FROM THE BRAIN, by S. Goldberg
PREPARING FOR MEDICAL PRACTICE MADE RIDICULOUSLY SIMPLE, by D.M. Lichtstein
MED'TOONS (260 humorous medical cartoons by the author) by S. Goldberg
CLINICAL RADIOLOGY MADE RIDICULOUSLY SIMPLE, by H. Ouellette
NCLEX-RN MADE RIDICULOUSLY SIMPLE, by A. Carl
THE PRACTITIONER'S POCKET PAL: ULTRA RAPID MEDICAL REFERENCE, by J. Hancock
ORGANIC CHEMISTRY MADE RIDICULOUSLY SIMPLE, by G.A. Davis
CLINICAL CARDIOLOGY MADE RIDICULOUSLY SIMPLE, by M.A. Chizner
PSYCHIATRY ROUNDS: PRACTICAL SOLUTIONS TO CLINICAL CHALLENGES, by N.A. Vaidya
        and M.A. Taylor.
MEDMASTER'S MEDSEARCHER, by S. Goldberg
PATHOLOGY MADE RIDICULOUSLY SIMPLE, by A. Zaher
CLINICAL PATHOPHYSIOLOGY MADE RIDICULOUSLY SIMPLE, by A. Berkowitz
ATLAS OF MICROBIOLOGY, by S. Goldberg
ATLAS OF DERMATOLOGY, by S. Goldberg and B. Galitzer

Try your bookstore. For further information and ordering send for the MedMaster catalog at MedMaster, P.O. Box 640028, Miami FL 33164. Or see http://www.medmaster.net for current information. Email: mmbks@aol.com.